Necropsy Techniques for Examining Wildlife Samples

Authored by

Andreia Garcês

CITAB, University of Trás-os-Montes and Alto Douro
Vila Real, Portugal
INNO – Veterinary Services, Braga
Portugal

&

Isabel Pires

CECAV, Department of Veterinary Sciences
University of Trás-os-Montes e Alto Douro
5000 Vila Real
Portugal

Necropsy Techniques for Examining Wildlife Samples

Authors: Andreia Garcês and Isabel Pires

ISBN (Online): 978-981-14-6833-9

ISBN (Print): 978-981-14-6831-5

ISBN (Paperback): 978-981-14-6832-2

need for a court order if at any point you breach any terms of this License Agreement. In no event will any delay or failure by Bentham Science Publishers in enforcing your compliance with this License Agreement constitute a waiver of any of its rights.

3. You acknowledge that you have read this License Agreement, and agree to be bound by its terms and conditions. To the extent that any other terms and conditions presented on any website of Bentham Science Publishers conflict with, or are inconsistent with, the terms and conditions set out in this License Agreement, you acknowledge that the terms and conditions set out in this License Agreement shall prevail.

Bentham Science Publishers Pte. Ltd.
80 Robinson Road #02-00
Singapore 068898
Singapore
Email: subscriptions@benthamscience.net

BENTHAM SCIENCE

CONTENTS

PREFACE

Post-mortem examination is an essential tool for determining the cause and circumstances of death. Even in the era of molecular pathology, necropsy remains the most valuable tool for understanding the whole organism and the disease.

In wildlife, the knowledge obtained from necropsy is much more comprehensive not only in pathology but also in several areas of biology (*e.g.* virology, microbiology, genetic). The correct interpretation of the phenomena surrounding the death can contribute to the identification of new diseases, re-emerging diseases, to the preservation of wildlife by identifying risk factors and threats to species survival. This should always be applied in a global health context.

One of the critical points of wildlife necropsy is the knowledge of anatomy, physiology and pathology of the different classes, and ultimately of a particular species. A necropsy of a different species is always challenging. Therefore, we try to apply and teach students of veterinary medicine and biology, some basic assumptions for maximizing wildlife necropsy success:

- Death is not the end, but a new path to generate knowledge that is essential for the preservation of life itself. The collection of samples for further exams is fundamental.
- It is necessary to look for the unity in the diversity of species, lesions and diseases.
- It is required to find the diversity of lesions in the unity of one etiological agent.
- Different lesions could represent different expressions of the same disease in different species.
- Similar macroscopic features could represent lesions or non-lesions according to the animal species.
- The death of each animal has its language. The challenge of every *post-mortem* exam is the comprehension of the message that each cadaver can transmit to us.
- The necropsy of wildlife takes us far beyond an individual diagnosis. The correct interpretation of lesions can be applied in the study of populations and ecosystem health.

In this publication, the authors aim to provide a practical, easily accessible guide of necropsy techniques in wildlife, addressing briefly some of its peculiar characteristics. It is mainly intended for students and professionals of biology and veterinary medicine areas. Being easy to consult, it also intends to be an auxiliary to professionals who work in natural parks, wildlife rehabilitation centres, biological or zoological parks, alerting to the importance of necropsy in the wildlife.

More than an anatomoclinic necropsy, with the objective of reach a diagnosis, the necropsy in wildlife is a unique moment to understand not only the death but also the life of the dead animal and its species and the health of its ecosystem.

To make the necropsy technique more didactic, in addition to a great diversity of animal images and techniques, schemes designed by one of the authors (Andreia Garcês) were provided.

Andreia Garcês
CITAB, University of Trás-os-Montes and Alto Douro
Vila Real, Portugal;
INNO – Veterinary Services, Braga
Portugal

&

Isabel Pires
CECAV, Department of Veterinary Sciences
University of Trás-os-Montes e Alto Douro
5000 Vila Real
Portugal

CHAPTER 1

Necropsy in Wildlife

Outline: In this chapter, we express the importance of the *post-mortem* examination in wildlife conservation. The authors also describe the necessary equipment, the difficulties presented, and the samples that can be obtained from this procedure.

Keywords: Animals Sentinels, Biodiversity, Forensic, Mortality, Necropsy, *Post-Mortem*, Wildlife.

THE IMPORTANCE OF *POST-MORTEM* EXAMINATION IN THE WILDLIFE

A wild animal, as defined by OIE, is an animal that has a phenotype unaffected by human selection and lives independent of direct human supervision or controls. Wildlife species exist in nature including the exotic and wild animals raised in captivity (*e.g.* zoos, parks, sanctuaries) sharing the same genetic code with its common ancestor. Although they may have been tamed, this characteristic is not transmitted to the next generation, and for that reason, they cannot fail to be considered as savages (World Organisation for Animal Health, 2010).

Post mortem examination is useful to assess the cause and death circumstances. However, in wild animals, necropsy is of additional importance. On one hand, it is a unique opportunity to study the anatomy, physiology, behaviour, habitat, and thus, increase the knowledge about the species. On the other hand, the necropsy of a wild animal does not an end in itself.

The *post mortem* findings could give accurate information about animal species, habitat, hazards to which it has been exposed, or even could contribute to the identification of areas and species at risk and priority intervention measures (King *et al.*, 2014; Terio, Macloose and Leger, 2018).

Necropsy in wild species may be more difficult than in the domestic species due to the variability of animal anatomy. Thus, the experience of the operator is fundamental. Besides knowledge of anatomy, physiology, and pathology, knowledge related to the biology of the species is also essential (Mörner, Obendorf and Al., 2002).

Thus, in addition to identifying the cause of death, additional measures may also be required for the endangered species, and for the notification to the competent authorities. Whenever it is possible, *post mortem* examination should be carried out on all corpses of wild animals found in the field, carcasses from hunting game origin, or animals dying in Wildlife Recovery Centres or Wildlife Sanctuaries. Even if the cause of death has already been determined (*e.g.* firearms, trampling), it is an indispensable exam capable of providing a vast amount of information at various levels, as well as allowing the collection of material for further examinations and understanding (Sainsbury *et al.*, 2001; Casal, Darwich and Molina-lo, 2013; Mullineaux, 2014).

Fig. (1). Diagram with some of the information that is possible to retreat during a *post mortem* exam including sample collection for genetic, parasitic, microbiological tests.

The necropsy should be considered as a privileged moment to study the anatomical, physiological, genetic, behavioural, and dietetic characteristics of the species as well as the microbial flora, parasites, pathological agents and the lesions that characterize the different diseases that affect those animals (Fig. **1**). At the same time, if possible, it is combined with other complementary tests such as radiography, computed tomography (CT), ultrasound, microbiological, parasitological, genetic, virologic and toxicological examination, to obtain as much information as possible (Friend and Franson, 1999; Simpson, 2000).

It is also an essential exam to understand the pathology of these animals, to identify new diseases, their aetiology, pathogenesis and evolution, as well as to evaluate possible treatments. In the study of outbreaks of infectious diseases, especially in epidemics in a given population or area, it sometimes becomes essential, not only to examine animals whose natural death occurred but also to the sacrificed animals that are affected with the disease for a complete evaluation of the nosological entity (Friend and Franson, 1999).

Humans' action on the ecosystem has an essential effect on the animal population and humans themself. Close contact with wildlife populations, which have often adapted to life within cities, increases the risk of disease transmission to humans and domestic animals. In this context, the role of the *post-mortem* examination of wild species plays a crucial role in the evaluation of zoonoses, making it possible to collect material for further examination (Friend and Franson, 1999; King *et al.*, 2013; Zachary, 2016).

It is thought that more than 70% of the currently emerging zoonoses have been originated from wild animals. The increase in these diseases is due to several factors, such as overexploitation of forests, increase areas for agriculture, illegal trade of wild animals, ecotourism in remote regions, wild game meat market, and others (Ditchkoff, Saalfeld and Gibson, 2006; Bradley and Altizer, 2007; Brearley *et al.*, 2013). These anthropogenic pressure on wildlife is leading to an imbalance of the ecosystems, with considerable losses not only in the terms of animal and human health but also in economic aspects.

FORENSIC WILDLIFE NECROPSY

While human threats to wildlife are increasing, there is also a growing concern about the welfare and conservation of these animals. Thus, cases of crimes against wildlife are frequently increasing.

There are several threats to wildlife, including illegal trade in animals (*e.g.* use of skin, bone or organs in alternative medicines, or ornament such as ivory), increased illegal poaching for bushmeat, kill of scavengers or predator's species with poison, anthropogenic threats (*e.g.* electrocution, collision with vehicles, collision infrastructures) or pollution (Wilson-Wilde, 2010; Huffman and Wallace, 2012).

Necropsy, as the ultimate tool for diagnosis, plays a fundamental role in the evaluation of the lesions in the case of suspected crime against wildlife. However, it will have medical-legal value if it is carried out in a detailed and correct manner, to maintain the chain of custody, in order to be able to hold the responsible culprits (Merck, 2008).

Therefore, the corpse must be collected by the responsible authorities, correctly identified and placed inside a sealed bag with the identification of the animal, keeper responsible for the collection, date and location. The agents of authority must prepare the accident report and collect as much data and samples on the spot as possible. The animal should then be sent to a necropsy lab, refrigerated or frozen to delay the phenomena of autolysis.

The primary purpose of forensic necropsy in wild animals is to determine the cause and circumstances of the animal's death. The identification of the species, sex, age is essential data, as well as the data of *ante mortem* history, data related to the environment and territory in which the animal lived, previous health status, behaviours, and events that may have influenced the cause of death is also required. With the information collected, the veterinarian can try to determine the chronological sequence of events and the possible cause of death, relying on the observed lesions. Sampling is also essential, *e.g.* samples for toxicology examination when poisoning is suspected (Fig. **2**) (Strafuss, 1988; Linacre, 2009; Bradley-Siemens and Brower, 2016; McDonough and McEwen, 2016; Brooks, 2018a, 2018b).

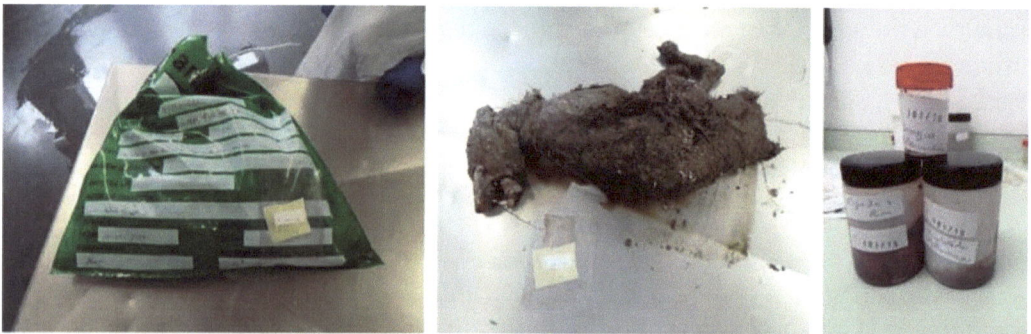

Fig. (2). Forensic necropsy to a *Vulpes vulpes* with suspicion of poisoning. Reception of the dead animal in the sealed bag and sampling for toxicology.

The veterinarian, preferably with experience in the field of pathology, should perform the necropsy systematically to collect as much information as possible. Anything that is observed should be recorded in the report, including samples that are collected, with or without lesions (for toxicological, histological examination). The samples should be appropriately identified and collected in duplicate. Photographs should be taken, preferably in formats that cannot be manipulated with computer programs. All photos should have a small paper with the identification of the case, animal, date, pathologist and organ. Rulings for lesion measurement should also be present.

After delivery of the report, it is the responsibility of the pathologist to keep copies of the report, photographs and samples in duplicate, until the judgment occurs or till the time determined by law (in some cases, it is mandatory to have the data stored for five years) (Linacre, 2009; McDonough and McEwen, 2016; Brooks, 2018a).

RISKS ASSOCIATED WITH WILDLIFE NECROPSY

The risks associated with wild animal necropsies are related to the transmission of diseases to humans, even when the animal is only the host of the infectious agent. Consequently, it is vital to fulfill all safety requirements during a necropsy to avoid contamination for those who handle the carcass and the environment.

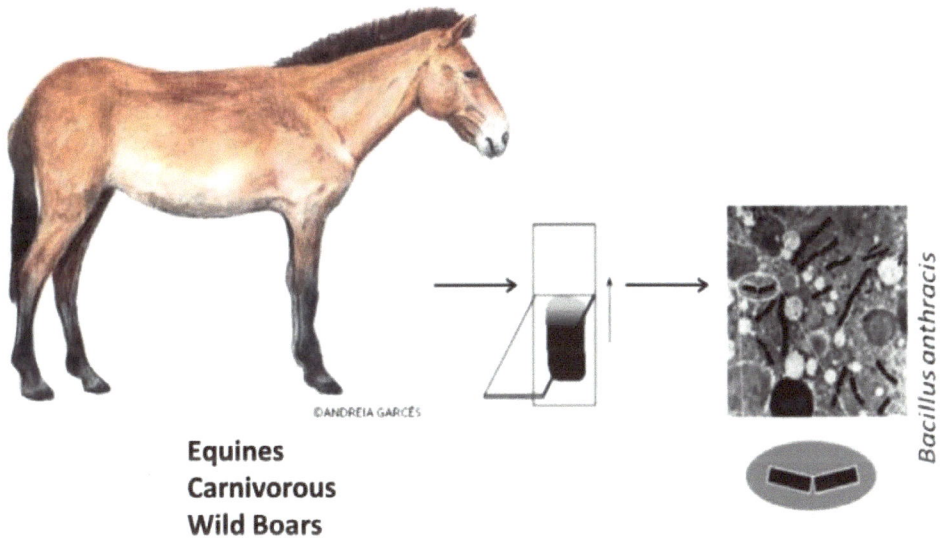

Equines
Carnivorous
Wild Boars

Bacillus anthracis

Fig. (3). Graphic representation of cytology for identification of *Bacillus anthracis* at 2000x magnification. The bacilli have a size up to 10 μm in length, a paler capsule, and a straight edge.

The technician who perform the procedure should be attentive if the cadaver has an inflated look with subcutaneous emphysema and a history of sudden death. In this case, it could be infected with *Bacillus anthracis,* particularly if it is a wild horse, ruminant, carnivore, or wild board (Figs. **3** and **4**). It is an extreme danger because if the cadaver is open it will expose the bacteria to the air and will form spores and disperse them into the environment, and will become deadly to every living thing in the area for long periods. The cadaver should not be open before despite the presence of this bacteria (Munson, 2000). In case of a positive diagnosis, the cadaver should be immediately eliminated by alerting the appropriate authorities to do so (burial in deep pits, more than 2 meters deep and covered with quicklime or other similar substance) (Munson, 2000; Woodford, Keet and Begins, 2000; Mörner, Obendorf and Al, 2002; Schmidt and Reavill, 2003).

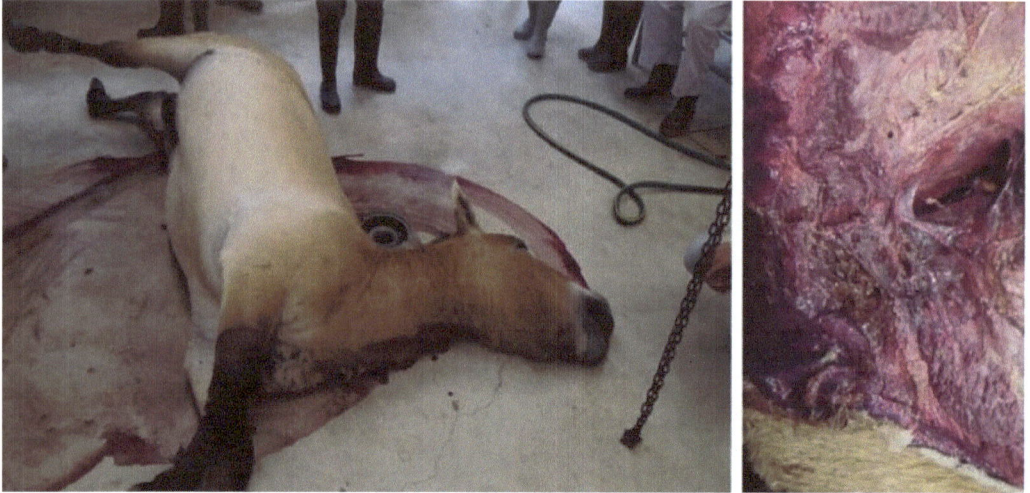

Fig. (4). Przewalski horse (*Equus ferus przewalskii*) with a suspect diagnosis of *Bacillus Anthracis.*

(Table **1**) lists some of the main zoonoses that affect the different classes of wild vertebrates, including diseases caused by viruses, bacteria and fungi.

In the specific case of non-human primates, their diseases represent a higher risk due to their phylogenetic proximity to humans. Many of them are very similar to the diseases that affect humans and are therefore easily transmissible to these and *vice versa*. Some examples of zoonoses transmitted by non-human primates are Herpesvirus type I, Hepatitis A (HAV), Rubella (virus), Leprosy (*Mycobacterium leprae*), Syphilis (*Treponema pallidum*), Dengue (*Flavivirus*: DEN- 2, DEN-3 and DEN-4), Monkeypox (*Orthopoxvirus*), Yellow Fever (*Flavivirus*), Ebola (*Ebolavirus*) and Marburg hemorrhagic fever (*Marburgvirus*).

Also, they have several immunodeficiency-associated retroviruses such as *Ortoretrovirus* type C, *Oncornaviruses* type D and Simian Immunodeficiency Viruses (SIV). SIVs are of great importance because they are human immunodeficiency virus-like lentiviruses, HIV-1 and HIV-2 (Munson, 2000; Woodford, Keet and Begins, 2000; Mörner, Obendorf and Al., 2002; Schmidt and Reavill, 2003).

Table 1. Brief list of some of the agents responsible for zoonoses present in wild animals.

	Mammals	Birds	Reptiles and Amphibians
Parasites	*Amblyomma americanum* *Giardia lamblia* *Toxoplasma gondii* *Ixodes scapularis* *Ixodes dammini* *Xenopsylla cheopsis* *Nasopsyllus fasciatus* *Ascaris lumbricoides* *Fasciola gigantica* *Necator americanus* *Ancylostoma duodenale* *Ancylostoma ceylanicum* *Trichuris trichuria* *Wuchereria bancrofti* *Brugia malayi* *Brugia timori* *Onchocerca volvulus* *Trypanosoma cruzi* *Trypaonsoma brucei gambiense* *Trypaonsoma brucei rhodisiense* *Leishmania infantum,* *Leishmania donovani* *Taenia solium* *Dracunculus medicinsis* *Paragonimus spp.* *Opisthorchis viverinni* *Opisthorchis felineus* *Sarcoptes scabiei* *Cryptosporidium spp.* *Baylisascaris spp.* *Echinococcus spp* *Rhipicephalus sanguineus* *Dermacentor variabilis* *Dermacentor andersoni* *Leptospira spp.*	*Giardia lamblia* *Giardia spp.* *Toxoplasma gondii* *Dermanyssus gallinae*	*Gnathostoma spp.*
Virus	*Rhabdovirus* *Arboviruses* *Hantavirus*	*PMV-1 infection* *Avian Influenza H5N1* *West Nile virus*	

(Table 1) cont.....

	Mammals	Birds	Reptiles and Amphibians
Bacteria	Brucella spp. Yersinia pestis Histoplasma capsulatum Borrelia burgdorferi Rickettsia rickettsi Salmonella spp. Clostriduim tetanii Eschericcia coli Mycoplasma sp. Bacillus anthracis Bartonella henselae Bartonella quintana Coxiella burnetii Afipia spp. Rickettsia akari Rickettsia typhi Erysipelothrix rhusiopathiae Listeria monocytogenes Mycobaterium leprae Leptospira spp.	Histoplasma capsulatum Clostriduim tetanii Salmonella spp. Bacillus anthracis Erysipelothrix rhusiopathiae Chlamydophila spp. Escherichia coli Yersinia enterocolitica Histoplasma capsulatum Listeria monocytogenes Pasteurella multocida Yersinia pseudotuberculosis Mycobacterium avium Bacillus anthracis	Salmonella spp. Aeromonas spp. Campylobacter spp. Klebsiella spp. Mycobacterium spp. Edwardsiella spp. Yersinia spp. Enterobacter spp. Flavobacterium meningosepticum Leptospira spp.
Fungi	Chlamydia trachomitis	Chlamydia psittaci Cryptosporidium spp. Aspergillus fumigatus	Zygomycosis Phycomycosis Mucormycosis Aspergillus spp. Candida spp. Cryptosporidium

ANTE-MORTEM HISTORY

Before initiating the necropsy, as much information as possible about the animal before its death should be collected: records of its behaviour, clinical signs, photographs, diet, places, climate, the hierarchy that occupies on the group (if they are among animals living in groups), and others.

In a healthy ecosystem, disease levels are deficient or go through periods of epidemics that help control and select populations in a balanced cycle. With the new threats posed by humans, from habitat destruction, pollution, hunting, illegal trade and climate change, this balance of epidemics is threatened.

Knowledge about the environment in which these animals are found is essential. It can provide us with several clues that can contribute to the understanding of the appearance of possible injuries and early diagnosis of the cause of death. In the ecosystem, abrupt climatic changes such as storms, intense precipitation, heat waves or extreme cold weather can lead to a decrease in food and water sources. These situations are a cause of stress, that can make the animals more susceptible to diseases. Knowledge about variations in water is crucial because they are often the cause of mortality, such as cases of botulism, cyanobacteria, toxic algae, petroleum spills, cleaning of ships, sewage, or pesticides. It is also essential to know regarding the food sources. If a decrease in food sources has occurred, the reasons for this decrease should be investigated (Strafuss, 1988).

It is still essential to know which transmitting disease vectors are present in the area of origin of the animal and how its population has been evolved. Also, information regarding the characteristics of the areas that were part of the territory of the animal, whether it includes urbanised or industrialised areas, for example, it is essential to determine the cause of death. With an excess of population in smaller spaces, the introduction of alien species into new territories, and an increase of illegal trade, it is a higher the risk of occurrence of epidemics, not only in wild populations but also in the human population (Simpson, 2000; Jakob-hoff *et al.*, 2014).

Some diseases are characteristic of certain species and others that affect a wide range of species. The age of the affected animals is also essential, since, associated with different diets and behaviours, the occurrence of the disease in the young or adults can contribute to its diagnosis.

The maximum amount of information related to the place where the corpse was collected should be retrieved, from the tracks of possible predators or scavengers, fluids present in the body, evidence of agonising death, baits, feather and hair, insects or faeces. The carcass may be originated from an animal that died acutely or chronically, only continues of the corpse that is found in the field, eggs/foetus or animals slaughtered by predators/hunters. It should be noted that in the case of groups or flocks of animals, it is sometimes necessary to sacrifice an animal to save the whole bunch.

All this information is essential and should be collected and transmitted to the pathologist who is responsible tp perform the *post-mortem* examination to improve the diagnosis.

NATURAL DEATH *VS.* EUTHANASIA

A vast majority of the animals on which the *post-mortem* examination is performed came from wildlife rehabilitation centres, zoos, sanctuaries or hospitals, and many of these animals were euthanised because their lesions were too severe to the appropriate animals' recovery and to be released back in the wild. The pathologist needs to know if the animal has been euthanised and which method was used. The different methods of sacrifice cause macroscopic and microscopic changes in the tissues, which should not be confused with lesions (Mullineaux, 2014).

Multiple laws (national and international) are responsible for the regulation of euthanasia of wildlife. It is a delicate subject matter since several species nowadays are protected and managed by several organizations and governments. Individual animals can be found injured, orphaned, or debilitated during the rehabilitation process and may require euthanasia when the animal cannot be returned to the wild and cannot be survive alone. The release of the individual would pose a threat to the health of the population in case of the inexistence of alternatives to the treatment and housing. In other cases, scientific research, population management, control of disease outbreaks, control of aggressive animals that are a nuisance to humans or collection protocols decree that some of these animals have to be killed (Leary *et al.*, 2020).

There are many factors to take into account when selecting the method of euthanasia for free-ranging or captive wildlife. Although some methods described for domestic animals may be useful for euthanizing free-ranging wildlife, their applicability is diverge. Also, due to their anatomical and physiological characteristic, some species may be too large to successfully euthanize (*e.g.* marine mammals) by the conventional means.

Another aspect is the potential for secondary toxicity and environmental hazards (scavenger exposure, lead poisoning) associated with the remains of animals euthanized (Leary *et al.*, 2020).

When possible, adequate physical or chemical restraint to allow the administration of the euthanasia should be used. When necessary, specific anaesthetics agents using remote injection equipment for chemical immobilization should be delivered if wildlife cannot be captured. The bibliography should be consulted to choose the correct doses of anaesthetics and anxiolytics and routes of administration particularly in animals that are kept in captivity (Leary *et al.*, 2020).

In general, the methods of euthanasia, according to the AVMA Guidelines for the Euthanasia of Animals (Leary *et al.*, 2020) that can be applied to wildlife are:

- **Chemical Methods:** overdoses of injectable anaesthetic agents (including barbiturates, potassium chloride). It is recommended to use the premedication with an injectable or inhaled agent to reduce animal distress (IV or IC).
- **Inhaled Anaesthetics, Carbon Dioxide, Carbon Monoxide, And Other Inert Gases:** are acceptable in small avian and mammalian species when other methods are not available.

- **Firearms:** is acceptable in free-ranging, captured, or confined wildlife, provided that bullet placement is on the head (targeted to destroy the brain). In the thorax (hearth) or to the neck (destroy the spinal cord) can be applied in free-ranging or other settings where the close approach is not possible or where the head must be preserved for disease testing (rabies, chronic wasting, or other).
- **Exsanguination:** is used as an adjunctive method and only applied after the animal is anaesthetized or unconscious.
- **Cervical Dislocation Or Decapitation:** used in small mammals and birds. This method may be useful as an adjunct or as a first-step method of euthanasia.
- **Thoracic Compression:** used only when the animals are deeply anaesthetized or unconscious, as a final or confirmatory method to ensure death.

In the next tables, some of the techniques are summarised which can be used for euthanasia with a brief description of the changes that can be observed in the *post-mortem* examination due to these procedures (Table **2**). It should be noted that the preservation method of the corpse influences these changes; for example, freezing makes the pentobarbital crystals of sodium cease to be visible (Friend and Franson, 1999).

Fig. (5). Different installations to perform the *post-mortem* exam according to the size of the animals and conditions available.

Next, we present the allowed methods of euthanasia in the different classes of vertebrates and invertebrates according to the AVMA Guidelines for the Euthanasia of Animals (Leary *et al.*, 2020), with some specification to the species.

Table 2. Summary of the physical methods of euthanasia, (P- small size animals, M - medium-sized animals, G – large size animals, T - all species) and post mortem findings related.

	Method	Description	Post-mortem Findings
PHYSICAL METHODS OF EUTHANASIA	Cervical dislocation	By force in opposite directions, a cervical displacement is made at the base of the brain or between the 1st and 2nd cervical vertebra, leading to the section of the bone marrow.	Section of the bone marrow Cervical stretch Cervical hematomas
	Decapitation	Section of the bone marrow with a sharp object (axe, knife)	Bruises Bleeding Continuity Solutions Bone marrow section Blood aspiration
	Stunning and exsanguination	Permanent numbness of the animal with a traumatism in the skull followed by exsanguination with the rupture of a large blood vessel by an ordinary cutting object.	Hematoma and cerebral oedema Haemorrhage and hematoma Continuity Solutions (the brain cannot be used later for exams) Ischemia tissues Blood Blood aspiration
	Fire gun	Shooting with a gun on the skull or neck.	Haemorrhage /Hematoma Projectiles Fractures
CHEMICAL METHODS OF EUTHANASIA	Inhalant anaesthetics (Halothane, IIsoflurane, Methoxyflurane, Enfluran)	Through masks or placing the animal in a container sealed with an anaesthetic.	Pulmonary Congestion Bruises Cyanosis
	Toxic gas (carbon monoxide, carbon dioxide)	Placement with a sealed container.	Hypoxia Cyanosis Pulmonary Congestion
	Lethal injection (Eutasil®, Eutasol®, Sodium Pentobarbital)	Intravenous administration according to the dose indicated in the package leaflet	Erythrolysis Oedema and coagulation changes in the lungs Deposits of white crystals on the surface of organs Obvious injection point

Amphibians and Reptiles

The euthanasia in these animals can sometimes be a challenge due to their high tolerance to hypoxia that can limit the effectiveness of methods based on anoxia and the difficulty to access the vasculature in some species. The confirmation of death in these animals can be a challenge since their hearts can beat even after brain death, and the presence of scales/bone plates in reptiles. Therefore, it is recommended to apply at least 2 euthanasia procedures at the same time. The animals should be restrained before the procedure. Physical restraining is possible for many species (particularly small and medium-size) with adequate equipment. Chemical restraint may be useful in some situations, particularly for venomous or large animals where human safety would be compromised by physical restraint. Chemical restraint at high doses may serve as a first or preparatory step of euthanasia in some situations.

- **Injectable Agents:** due to the anatomy of some species, intravenous access can be difficult. Intracoelomic, subcutaneous lymph spaces, and lymph sacs are suitable routes of administration. Direct injection into the brain through the parietal eye (always under anaesthesia) can be used in some lizard species. Sodium pentobarbital can be administered in many species. In poikilotherms, dissociative agents as ketamine hydrochloride, combinations such as tiletamine and zolazepam, inhaled agents, propofol, or other barbiturates, may be used.
- **External Or Topical Agents:** In amphibians, buffered MS 222 *via* water baths can be used. Also, benzocaine hydrochloride may be used as a bath or applied topically to the ventrum as a gel.
- **Inhaled Anaesthetics and Carbon Dioxide:** it should only be used as the last resource since many species can survive during prolonged periods of anoxia (some up to 27 hours). Death may not occur even with prolonged exposure, and therefore it must be verified before terminating the use of the inhaled agent and applied a second procedure (*e.g.*, decapitation) to ensure death. It is useful in small size species.
- **Gunshot:** can be applied to crocodilians and other large reptiles (delivered to the brain).
- **Manually Applied Blunt Force Trauma to the Head:** it is used as an adjunctive method, for decapitation or pithing. It should only be used when the other methods are not available.
- **Rapid Freezing:** reptiles and amphibians can be euthanized by rapid freezing when it results in immediate death (< 4 g (0.1 oz) in liquid N).
- **Decapitation:** should only be performed as part of a third step euthanasia protocol (injectable anaesthetics, decapitation, pithing). It should be completed using heavy shears or a guillotine.

- **Pithing:** it can be used as a second-step euthanasia method. The pithing site in frogs is the foramen magnum.

Invertebrates

In this group, the great dilemma is the ability to perceive pain in these animals. The euthanasia methods used should minimize the potential for pain or distress.

- **Injectable Agents:** overdose of pentobarbital or similar agent, on a weight-to-weight basis. Premedication with an injectable or inhaled agent may facilitate administration. They should be injected directly into the circulating haemolymph, when not possible is possible to use an intracoelomic injection.
- **Inhaled Anaesthetics:** overdose of these drugs is acceptable for terrestrial invertebrates where other methods are not available. Adjunctive method of euthanasia is recommended, to ensure death.
- **Carbon dioxide:** could be used in some terrestrial invertebrates.
- **Physical Methods (*e.g.*, Boiling, Freezing, Pithing)/ Chemical (*e.g.*, Alcohol, Formalin) Methods:** destroy the brain or major ganglia, and should be used after the anaesthesia methods. A two-step euthanasia procedure is acceptable, starting with an immersion in 5% laboratory-grade ethanol or an undiluted, uncarbonated beer (5% ethanol content) to anaesthetize and then immersion in solutions of 70% to 95% ethanol or neutral-buffered 10% formalin (very used in snails).
- **Pithing:** This method requires detailed anatomic knowledge of the species in question.

Fish and Aquatic Invertebrates

The effectiveness of euthanasia methods may vary by species and life stage. Early stages in the lives of fish, including embryos and larvae, may require higher concentrations of immersion anaesthetics or a longer duration of exposure. An overdose of anaesthetics is suitable to apply to aquatic invertebrates as it is for fish. In these animals use two-stage euthanasia, in step 1, immersion to render the fish unconscious, and in step 2, a secondary adjunctive method to complete euthanasia (such as decapitation, pithing, or freezing).

- **Immersion in Anaesthetics Solutions:** the animals should be immersed in the anaesthetic's solution for a minimum of 30 minutes after cessation of opercular movement to ensure death. Some immersion agents are: benzocaine or benzocaine hydrochloride (buffered), carbon dioxide, ethanol (10 to 30 mL of 95% ethanol/L), eugenol, isoflurane, sevoflurane (> 5 to 20 mL/L), quinaldine sulfate ($\geq$ 100 mg/L), tricaine methanesulfonate (buffered), 2-phenoxyethanol

($\geq$ 0.5 to 0.6 mL/L or 0.3 to 0.4 mg/L, lidocaine (400 mg/L).

- **Injectable Agents:** IV, IC, IM, and intracardiac routes. Some agents that can be used are: sodium pentobarbital (60 to 100 mg/kg), ketamine (66 to 88 mg/kg), ketamine-medetomidine (1 to 2 mg/kg -0.05 to 0.1 mg/kg), propofol (1.5 to 2.5 mg/kg).
- **Decapitation Or Cervical Transection Followed By Pithing:** should be the second step of euthanasia. The rapid severance of the head and brain from the spinal cord, followed by pithing of the brain, will cause rapid death.
- **Blunt Force Trauma (Cranial Trauma) Followed By Pithing Or Exsanguination:** a rapid, accurately placed blow to the cranium with an appropriate-sized club can cause immediate unconsciousness and potentially death but should be followed by pithing or exsanguination to ensure death. Anatomic features, such as the location of the eyes, can help serve as a guide to the location of the brain.
- **Penetrating Captive Bolt Or Npcb:** applied to large fish species.
- **Maceration:** use a macerator specifically designed for the size of fish being euthanized (very small animals), death is nearly instantaneous.
- **Rapid Freezing:** applied to zebrafish (*D rerio*), small-bodied (up to 3.8 cm long) tropical and subtropical endothermic fish. The animals are put in a bath with 2° to 4°C cold water until its loss of orientation and operculum. This method should not be used in temperate, cool, or cold-water–tolerant fish, such as carp, koi, goldfish, or other species that can survive at 4°C and below.
- **Marine mammals:** due to their size, the main difficulties are restraining and to determine when they are in conscience or death.
- **Injectable Agents** anaesthetics that can be used alone or in combination (*e.g.* tiletaminezolazepam, ketamine, xylazine, meperidine, fentanyl, midazolam, diazepam, butorphanol, acepromazine, barbiturates, etorphine), are safe and easy to use in small to medium-sized animals. Not advisable to use in large animals due to the difficulty in restraining them and due the to use of large amounts of the drug. Sometimes in cetaceans, peripheral vasoconstriction or hypovolemic shock may limit access to peripheral veins and fat layers must be bypassed for an effective administration. Routes mucocutaneous *via* the blowhole (the safest and effective method), IV, IP, IM, IC (anaesthetized, moribund, or unconscious animals when there is appropriated equipment). Potassium chloride or succinylcholine may be used in animals that are anaesthetized or unconscious, thought IV or IC, with a low risk of secondary toxicity for scavengers when it is not possible to dispose of the remains (*e.g.*, deep burial, rendering).
- **Gunshot** in small marine mammals when injectable methods are not practical (use in free-ranging, captured, or confined animals). Recommended for use in large odontocetes or large mysticetes. The bullet placement is to the head (targeted to destroy the brain). It should be targeted to the heart (chest) or to the

neck (vertebrae, with the intent of severing the spinal cord) where the close approach is not possible or where the head must be preserved for disease testing (rabies, chronic wasting, or others).

- **Manually Applied Blunt Force Trauma:** blow to the head for small juvenile marine mammals when no other method is available.
- **Implosive Decerebration:** decerebration of large mysticetes and odontocetes through the detonation of properly placed, shaped, and dimensioned explosive charges.
- **Inhaled Anaesthetics:** not the best method to use because marine mammals can breath-hold which means extended periods of physical restraint for their administration. Some examples of drugs used are halothane, isoflurane, sevoflurane, methoxyflurane, enflurane or carbon dioxide, carbon monoxide, and other inert gas This method should be preferably used in small animals that can be confined in enclosed containers. In larger animals when they are substantially debilitated, sedated, or anaesthetized.
- **Exsanguination:** used as an adjunctive method to ensure the death of animals that are anaesthetized or otherwise unconscious.
- **Cervical Dislocation and Decapitation:** can be used as the first-step and last, respectively, method of euthanasia.
- **Thoracic Compression** in animals that are deeply anaesthetized or otherwise unconscious, or as a final, as a confirmatory method.

Terrestrial Mammals

There is a large variety of anatomic, physiologic, behavioural, and size variations in this group that sometimes can pose a challenge in the method of euthanasia to apply. The restraint can be physical with nets and other equipment for small and medium-sized species. In larger and aggressive animals use chemical restraint with anaesthetics and/or anxiolytics *via* IM or IV.

- **Injectable Agents:** barbiturates (IV, IP, IC) have a rapid action and are applicable in a wide range of species Opioids and other anaesthetics may be administered IV or IM. Potassium chloride (IV or IC) can be used in deeply anaesthetized or unconscious animals adjunctively.
- **Inhaled Anaesthetics, Carbon Monoxide, Carbon Dioxide, And Inert Gases** may be administered *via* face mask or chambers. Very small animals can be placed entirely crate into a chamber.
- **Gunshot** may be appropriate for some species as a first step or adjunct method of euthanasia.
- **Thoracic Compression** is a useful method as confirmatory of death after the first procedure has been applied.
- **Exsanguination:** only used as a secondary or tertiary method to ensure death.

- **Cervical Dislocation Or Decapitation** can be used as an adjuvant to other methods in small mammals.

Birds

The choice of euthanasia method depends considerably on the species, size, anatomic and physiologic characteristics, clinical state, and response to restraint. Since birds lack a diaphragm (they have a single coelomic cavity) during the administration of intracoelomic injections the technician must be careful to not inject the drug into the air sacs, which could drown the bird or irritate the respiratory system. Also, euthanasia drugs must not be administered *via* the intraosseous route into the humerus or femur, because the hollow and pneumatic bones, communicate directly with the respiratory system, and drowning or irritation of the respiratory system may occur. Because of their greater capacity to process air, birds are more sensitive to inspired toxicants than other animals. Physical restraint is possible for many bird species, using equipment such as nets and trained personal.

- **Injectable Agents** are the quickest and most reliable means of euthanizing. Sedative or anaesthesia before the procedure. Barbiturates and barbituric acid derivatives are some of the drugs used.
- **Inhaled Anaesthetics:** used at high concentrations (*e.g.*, halothane, isoflurane, sevoflurane, with or without N_2O) as a sole method of euthanasia in particular cases, such as in large flocks or a large number of animals. It can be used in association with other methods.
- **Carbon Dioxide, Carbon Monoxide** only in high (> 40%) concentrations of CO_2 induce anaesthesia initially followed by loss of consciousness. Neonatal birds, because the unhatched bird's environment (with CO_2 concentrations as high as 14% in the embryonic chicken), require high concentrations to achieve euthanasia (to 80% to 90%). Also, diving species have physiologic adaptations to hypercapnia and may require higher concentrations.
- **Cervical Dislocation:** been used for small birds (< 200 g) when no other method is available.
- **Decapitation** in small (< 200 g) birds associated with other methods, after anaesthesia or unconscious animals.
- **Exsanguination** unconscious or anaesthetized birds. Useful blood samples for diagnostic or research purposes.
- **Thoracic Compression** is an adjunctive method in a second ou third step of euthanasia to ensure death.

MATERIAL AND EQUIPMENT

The necropsy can be performed both in the field and in a room suitable for this purpose. If it is done in the field, a prepared kit should be available. This kit should have, as basic material, cutting material such as knives, scalpels, scissors, forceps, forceps, axes and bone saws. This can be adapted to the size of the animal and the type of external revetment. This material must always be correctly identified as belonging to the necropsy service and should never be used for another purpose. It should always be cleaned and disinfected after each intervention (Fig. **5**).

As individual protection material, the kit must include latex gloves, masks, caps, foot protectors and a disposable gown. In case it is not possible to have disposable equipment, the clothes must be easily washable (King *et al.*, 2014).

It is also essential to always have material to write down the collected information such as notepad, audio recorder, camera, camera and necropsy protocols that contain all the following steps, so the technician does not lose any data.

The place where the necropsy is performed should be separated from the clinic area, without contact with other live animals. Infrastructures should be easily washable with adequate water supply, drainage and ventilation.

At the necropsy site, it is essential to have material for measurements such as rulers and scales.

Sampling material should not be overlooked, as described in the following topic and always have at hand material for transport and storage of samples such as ice chests, thermal boxes, card boxes, *etc.* (King *et al.*, 2014).

THE NECROPSY REPORT

The autopsy report is an official document, issued at the time of the necropsy that should include all the relevant observations. Your report should be simple, clear, direct and concise. The language should be correct, without stylistic features or regionalisms (Munson, 2000; Linacre, 2009; Somvanshi and Rao, 2009; King *et al.*, 2013; Brooks, 2018b).

It should include a description of the lesions and, where possible, the interpretation of the lesions and be completed by a macroscopic diagnosis.

The report must follow a logical order, generally consistent with the order of necropsy. After identification of the animal and clinical history, if it exists, it should include the animal weight and the description of their body condition.

Following the description of the internal and internal habit is the macroscopic diagnosis. It should also mention any sample collection for other types of exams.

The description of the lesions should include their identification and characterisation of their location, shape, distribution, size, colour, consistency and texture and other special features (Strafuss, 1988; Munson, 2000; Hocken, 2002; Schmidt and Reavill, 2003; Linacre, 2009; Somvanshi and Rao, 2009; King *et al.*, 2013; Brooks, 2018b).

REFERENCES

Bradley-Siemens, N & Brower, AI (2016) Veterinary Forensics: Firearms and Investigation of Projectile Injury. *Vet Pathol,* 53, 988-1000.
[http://dx.doi.org/10.1177/0300985816653170] [PMID: 27312366]

Bradley, CA & Altizer, S (2007) Urbanization and the ecology of wildlife diseases. *Trends Ecol Evol (Amst),* 22, 95-102.
[http://dx.doi.org/10.1016/j.tree.2006.11.001] [PMID: 17113678]

Brearley, G, Rhodes, J, Bradley, A, Baxter, G, Seabrook, L, Lunney, D, Liu, Y & McAlpine, C (2013) Wildlife disease prevalence in human-modified landscapes. *Biol Rev Camb Philos Soc,* 88, 427-42.
[http://dx.doi.org/10.1111/brv.12009] [PMID: 23279314]

Brooks, JW (2018) *Veterinary Forensic Pathology.* Springer International Publishing, NY, USA.

Brooks, JW (2018) *Veterinary Forensic Pathology.* Springer International Publishing, NY, USA.

Molina-López, RA, Casal, J & Darwich, L (2013) Final disposition and quality auditing of the rehabilitation process in wild raptors admitted to a Wildlife Rehabilitation Centre in Catalonia, Spain, during a twelve year period (1995-2007). *PLoS One,* 8, e60242.
[http://dx.doi.org/10.1371/journal.pone.0060242] [PMID: 23613722]

Ditchkoff, SS, Saalfeld, ST & Gibson, CJ (2006) Animal behavior in urban ecosystems: Modifications due to human-induced stress. *Urban Ecosyst,* 9, 5-12.
[http://dx.doi.org/10.1007/s11252-006-3262-3]

Friend, M & Franson, JC (1999) *Field Manual of Wildlife Diseases - General Field Procedures and Diseases of Birds.* Library of Congress, Cataloging, USA.

Huffman, J & Wallace, J (2012) *Wildlife forensics: Methods and Applications.* Wiley-Blackwell, West Sussex.

Jakob-Hoff, RM (2014) *Manual of Procedures for Wildlife Disease Risk Analysis* OIE, Paris 163.

King, JM (2013) *The necropsy book: A Guide for Veterinary Students, Residents, Clinicians, Pathologists, and Biological Researchers.* College of Veterinary Medicine, Cornell, University, NY, USA.

King, JM 2014 *The necropsy book: A Guide for Veterinary Students, by The Necropsy Book.* Charles Louis David DVM Foundation Publisher, Ithaca.

Leary, S (2020) *AVMA Guidelines for the Euthanasia of Animals.* American Veterinary Medical Association, Schaumburg.

Linacre, A (2009). *Forensic Science in Wildlife Investigations.* CRC Press Taylor& Francis Group, Boca Raton, USA.
[http://dx.doi.org/10.1201/9780849304118]

McDonough, SP & McEwen, BJ (2016) Veterinary Forensic Pathology: The Search for Truth. *Vet Pathol,* 53, 875-7.
[http://dx.doi.org/10.1177/0300985816647450] [PMID: 27515387]

Merck, MD (2008) *Veterinary Forensics: Animal Cruelty Investigations.* Academic Press, UK.

Mörner, T, Obendorf, DL, Artois, M & Woodford, MH (2002) Surveillance and monitoring of wildlife diseases. *Rev - Off Int Epizoot,* 21, 67-76.
[http://dx.doi.org/10.20506/rst.21.1.1321] [PMID: 11974631]

Mullineaux, E (2014) *Veterinary treatment and rehabilitation of indigenous wildlife.* Journal of Small Animal Practice, 55, 293-300.
[http://dx.doi.org/10.1111/jsap.12213]

Munson, L *Necropsy of Wild Animals.* Wildlife Health Center, School of Veterinary Medicine, USA.

Somvanshi, R & Rao, JR (2009) *Necropsy techniques and necropsy conference manual.* Veterinary Research Institute, India.

Sainsbury, AW, Kirkwood, JK, Bennett, PM & Cunningham, AA (2001) Status of wildlife health monitoring in the United Kingdom. *Vet Rec,* 148, 558-63.
[http://dx.doi.org/10.1136/vr.148.18.558] [PMID: 11370880]

Schmidt, RE & Reavill, DR (2003) *A Practitioner's Guide to Avian Necropsy.* Zoological Education Network, Lake Worth, Florida.

Simpson, VR (2000) Veterinary advances in the investigation of wildlife diseases in Britain. *Res Vet Sci,* 69, 11-6.
[http://dx.doi.org/10.1053/rvsc.2000.0384] [PMID: 10924388]

Strafuss, AC. (1988) *Necropsy Procedures and Basic Diagnostic Methods for Practicing Veterinarians.* Charles C. Thomas, Springfield, Illinois.
[http://dx.doi.org/10.1111/j.1939-165X.1988.tb00498.x]

Terio, KA, Macloose, D, St. Leger, J (2018) *Pathology of Wildlife and Zoo Animals.* Academic Press, UK.

Wilson-Wilde, L (2010) Wildlife crime: a global problem. *Forensic Sci Med Pathol,* 6, 221-2.
[http://dx.doi.org/10.1007/s12024-010-9167-8] [PMID: 20512431]

Woodford, MH, Keet, DF & Begins, RG (2000) *Post-mortem procedures for wildlife veterinarians and Fiel Biologists.* Iucn, Paris, France.

World Organisation for Animal Health. (2010) http://www.oie.int/en/

Zachary, JF (2016) *Pathologic basis of veterinary disease.* Elsevier, USA.

CHAPTER 2

Sample Collection

Outline: In this chapter, we describe the samples that can be collected from a cadaver in the field and during the *post mortem* exam.

Keywords: Corpse, Necropsy, *Post-Mortem*, Samples, Wildlife.

SAMPLING

A post-mortem examination in wildlife allows diagnosis of the cause of death and the detection of subclinical disease. Nevertheless, it should complement a full range of supporting tests (Cooper and Cooper, 2013). There are many and different diagnostic tests available. Even in the absence of gross lesions, it is important to collect samples to understand what is normal in these animals, in which little is known.

If possible, samples must be collected also when the animals are still alive. Some of the most important laboratory tests in the diagnosis of wildlife disease are histopathology, microbiology, parasitology, serology, among others (*e.g.* electron microscopy, toxicology, cytology, genetic, *etc.*) (Munson, 2000; Survey, 2000; Somvanshi and Rao, 2009) (Fig. **1**).

It is important to follow a consistent method, in order to produce consistent results and decrease the probability of error at this stage. This is vital when investigating an epizootic or attempting to establish the health status of a population (Cooper and Cooper, 2013). However, when carrying out *post-mortem* examinations in the field, some tests are almost impracticable because of a lack of facilities or likely delay in getting samples to the laboratory. Sometimes, the use of kits can be helpful. Nevertheless, the results obtained still will depend on the quality of the sample (Van Bressem, Van Waerebeek and Raga, 1999; Cooper, 2002; Cooper and Cooper, 2013).

Andreia Garcês & Isabel Pires

Fig. (1). Examples of different types of samples that can be taken from a cadaver during *post-mortem* examination and material – microbiology, serology, cytology, toxicology, parasitology.

Next, we describe the most common methods of sampling:

HISTOLOGY

For histopathological analysis, samples of tissue should be from all organs, including those in which no gross lesions are observed (Flint *et al.*, 2009).

Samples should be collected from relatively fresh carcasses, and never from frozen carcasses. Before collecting samples, do the external and internal evaluation of the carcass making a writing registry and photographing, if possible, any abnormality present (White and Dusek, 2015; Franson, 2016). Multiple samples should be taken from the larger organs, from representative sites. Sampling should include, when possible, the different anatomical constituents of each organ.

The transition between the lesion and normal tissue should preferably be collected. In the case of large samples, since it is impossible to send the lesion in its entirety, its different aspects should be sent different cuts representative of the lesion. Smaller specimens or specific lesions may be placed directly into cassettes and fixed (Munson, 2000; Coles, 2007) (Fig. **2**). Samples should not be crushed with forceps, and the mucosa of the gastrointestinal tract or the urinary bladder should not be handled or washed with water. Tissue collected should not exceed 5 mm in thickness, the samples should be 20 x 20 x 3 mm in size. Expect in the cases of spinal cord, eye and brain, which are collected and fixed whole (Flint *et al.*, 2009).

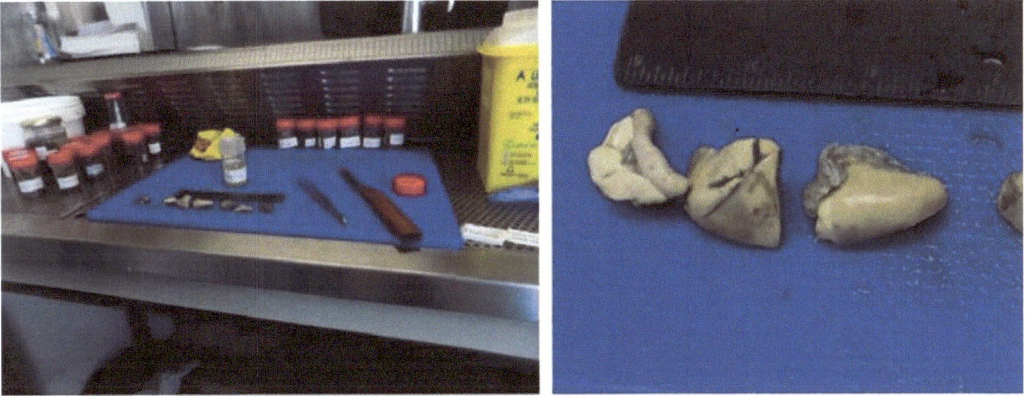

Fig. (2). Tissue samples preserved in 10% neutral formalin for histopathological examination.

If samples were very thick, an incision should be made so that the fixative solution to ensure maximum penetration into the tissue. In the case of hollow organs, such as the intestine, the fixative can be inoculated to the lumen. The lung, due to a large amount of air, should be fixed by placing a small veil on the surface. The brain and eye should be fixed in their totality (King *et al.*, 2014). The sample must be wholly immersed in a fixative solution (about ten times the volume of fixative concerning the sample).

The most widely used fixative for tissue samples is 10% neutral buffered formalin. If 10% neutral buffered formalin is not available, pure formalin (formaldehyde solution at 37-40%) may be mixed with filtered distilled water at a change to 10 parts pure formalin to 90 parts distilled water (10:90). Fixation should be complete in one to two weeks (Flint *et al.*, 2009). It is advisable that the formalin is changed 24 hours post collection, and verified routinely during storage to avoid loss of fluid. Formalin is a hazardous and potentially cancerigenous solution; and therefore, appropriate measures should be taken to prevent skin contact or vapour inhalation. Wide-mouth plastic bottles are the ideal contender to be used as containers for the preservation of tissues (White and Dusek, 2015).

MICROBIOLOGY

To collect material for microbiological analysis, sterilised equipment must always be used. As a base material should be present in the room sterile syringes and needles, swabs, tubes with culture medium, glass tubes, sterile Petri dishes, flasks, Eppendorf's, slides and coverslips for smearing or cytology. There should preferably be a nearby flame source, such as a lamp or lighter, to render the field as sterile as possible in which the operation will be carried out and to avoid contamination. Always use clean material (ex. scissors, scalp) when collecting samples for microbiology.

Specimens can be taken from skin, internal organs, and fluids may be taken for isolation and identification of bacteria, viruses, fungi and pathogenic protozoa. Samples may be stored in the appropriated medium (ex. agar). Swabs are an extremely valuable tool for sampling dead and live animals. Transport swabs should be removed from their sterile wrapping, brushed against the lesion or fluid requiring sampling (Fig. **3A**). Tracheal (typically used on dead birds), oral pharyngeal, cloacal/anus, and nasal swabs are some of frequently used. Also, faecal and environmental swabs can be used to examine the persistence of a determined pathogen in the environment or when it is not possible to perform directly a swab in the animal.

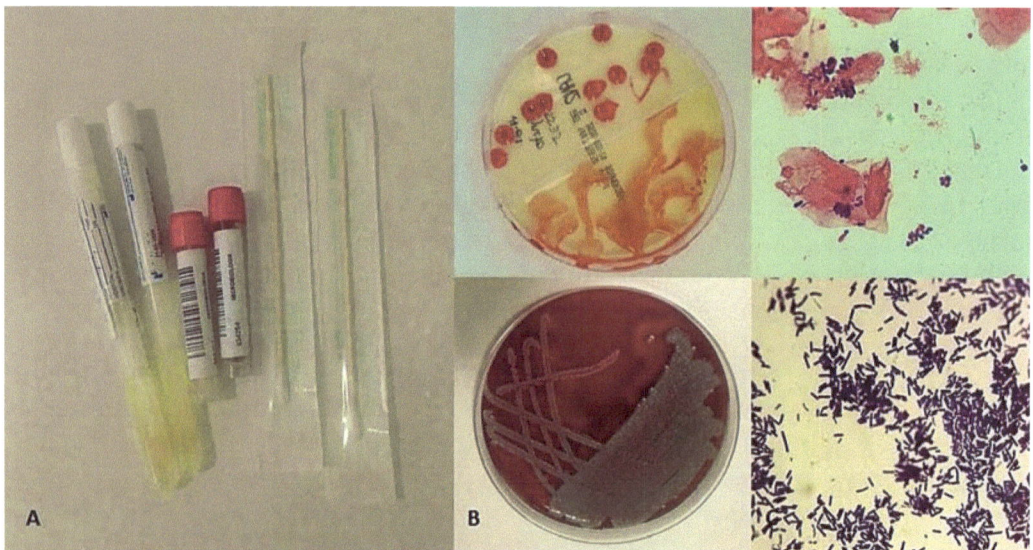

Fig. (3). Examples of some of the material used in microbiology. **A:** A transport swabs; **B:** culture mediums; **C:** gram colouration of yeasts and gram-positive bacillus.

Samples should be stored in a cool, dry environment. In the case of serum, blood should be collected into a clot tube and frozen at long term if required at -20° C. cytology's by aspiration or printing of tissue of fluids can ve perform and routinely coloured with Gram, Ziehl-Nielsen, Diff-Quick or Lactophenol to observe the presence of structures compatible with fungus ou microorganisms (Figs. **3B & C**). They can be performed after the preparation is dry and observed under a microscope or sent to the laboratory without colouring. The sample should be processed as soon as possible after collection to prevent degradation and contamination, within 72 hours (Friend and Franson, 1999; Flint *et al.*, 2009; White and Dusek, 2015). Haemolysed blood from fresh carcasses can be collected from the heart and stored frozen for virology.

PARASITOLOGY

All parasites found both internal and external should be collected. Nematodes, trematodes and cestodes should first be dropped into water or glacial acetic acid for one minute to dir. Them transferring them to 80% alcohol for storage. Flukes may be killed and relaxed using hot (80°C) water. Larger worms should then be placed on a piece of moist filter paper in a petri dish and another piece of filter paper should be placed on top with 10% formalin solution added, drop-by-drop until the filter paper is damp. They become fixed after one hour (Eros *et al.*, 2007). Ectoparasites and eggs from the same host may be preserved in a single vial in 70% alcohol (Farris *et al.*, 2014). Protozoan parasites on "fresh *postmortem* faecal samples" through faecal flotation or faecal smears (Fig. **4**) (Friend and Franson, 1999).

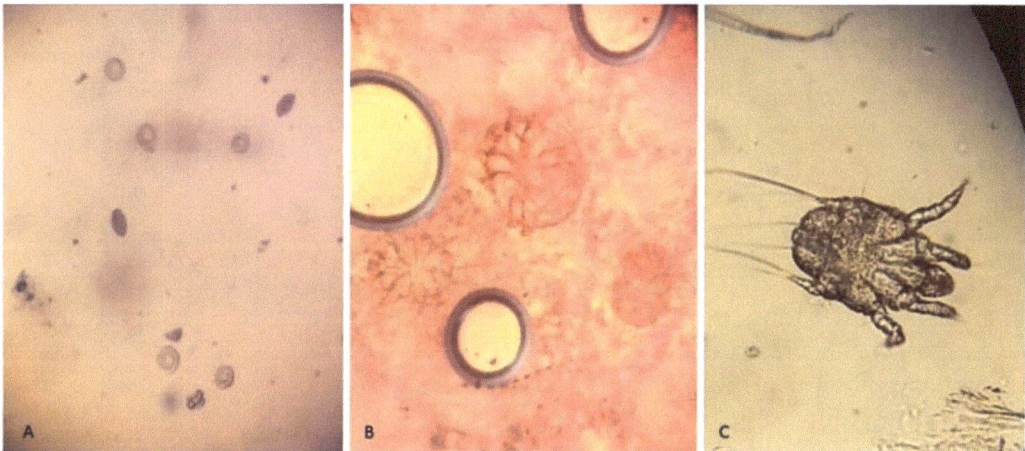

Fig. (4). Parasites observed with Willis method **(A)** in faeces and ectoparasites in otitis by cytology **(B, C)**.

GENETIC

Blood, muscle, gonad or skin samples may be taken for genetic studies, preferably from a fresh carcass. Samples may be collected in a variety of methods. Skin and musk may be collected by using a biopsy punch or scalpel to remove a 5 - 10 mm section of and be stored in a NaCl-saturated solution of 20% DMSO. If collected blood, 10 mL of blood (depending on the size of the animal) may be collected *via* cardiac puncture to be frozen or stored in 100 mmol Tris, 100 mmol NaCl, 10 mmol EDTA.2Na, 0.5% SDS). In non-fresh carcass, the skin and/or muscle is the most favoured tissue to collect. A sample of 1 cm x 0.5 cm should be collected and preserved frozen, in 80% EtOH, 20% DMSO in saturated NaCl solution or dried (Eros *et al.*, 2007; Flint *et al.*, 2009).

Another material that can be used for genetic tests is faecal samples. The faeces should be collected from the distal colon of a fresh. They should be preserved by frozen (at -80°C), storage in 80% EtOH, 0.01M Phosphate Buffered Saline (NaCl 0.138 M; KCl 0.0027 M) or 20% DMSO in saturated NaCl solution (Friend and Franson, 1999; Eros *et al.*, 2007; Farris *et al.*, 2014).

TOXICOLOGY

Specimens from the skin, adipose tissue, kidney, liver, muscle, urine, bone may be collected for the toxicological exam to access levels of toxins, heavy metals, and pollutes. With a clean blade (water and ethanol), even between sampling different tissues collect the samples. Approximately 2 g of tissue is sufficient for toxicology tests, but 5–10 g may be required if multiple toxicology tests will be completed (Munson, 2000; Eros *et al.*, 2007; White and Dusek, 2015).

Samples should be placed individually in glass jars and frozen as soon as possible after collecting. If not available, samples taken for heavy metal analysis should be frozen and placed in separate plastic bags. Samples taken for pesticide analysis should be frozen and wrapped in aluminium foil (Eros *et al.*, 2007; Flint *et al.*, 2009).

CONCLUSIONS

Samples that have been collected must be duly identified with the necessary data such as identification of the animal, species, age, sex, date, name of who performed the necropsy and what organs are present. These samples must be accompanied by the necropsy report duly completed and with *ant-mortem* history, whenever possible (Schmidt and Reavill, 2003).

If the samples need to be transported to the laboratories they should be packaged in heavy-duty containers and should comply with the appropriate local protocol. Specimens sent in bottles should be sealed with tape and enclosed in sealed plastic bags, and they put in a strong insulated container or cardboard box. Frozen samples must be transported in very sturdy ice chests, using ice bricks to ensure that its stay cold (Eros *et al.*, 2007).

The carcass should be disposed of in the end to prevent the spread of disease agents to other animals through environmental contamination. Incineration is the main method of disposal. The equipment and environment also should be clean, using a 10 per cent (1 part bleach to 9 parts water), due to its effectiveness (10 min contact), availability, and rapid decomposition in the environment (White and Dusek, 2015).

REFERENCES

Van Bressem, MF, Van Waerebeek, K & Raga, JA (1999) A review of virus infections of cetaceans and the potential impact of morbilliviruses, poxviruses and papillomaviruses on host population dynamics. *Dis Aquat Organ,* 38, 53-65.
[http://dx.doi.org/10.3354/dao038053] [PMID: 10590929]

Coles, B (2007) Post-mortem Examination. *Essential of avian medicine and surgery* Blackwell, London 103-15.

Cooper, J & Cooper, ME (2013) *Wildlife forensic investigation: Principles and Practice.* CRC Press Taylor & Francis Group, Boca Raton.
[http://dx.doi.org/10.1201/b14553]

Cooper, JE (2002) Diagnostic pathology of selected diseases in wildlife. *Rev - Off Int Epizoot,* 21, 77-89.
[http://dx.doi.org/10.20506/rst.21.1.1320] [PMID: 11974632]

Eros, C (2007) *Procedures for the Salvage and Necropsy of the Dugong (Dugong dugon).* Great Barrier Reef Marine Park Authority, Australia 104.

Farris, SC (2014) *Necropsies of Reptiles : Recommendations and Techniques for examining Invasive Species.* IFAS Extension University of Florida, Florida 26.

Flint, M (2009) Recording and submitting specimen history data. *Field manual of wildlife diseases, US Geological Survey, US Fish and Wildlife Service, and National Park Service* 3-6.

Bodenstein, BL (2016) Recording and Submitting Specimen History Data In: Franson, J.C., Friend, M., Gibbs, S.E.J., Wild, MA., (Eds.), *Field Manual of Wildlife Diseases.*U.S. Geological Survey Techniques and Methods. book 15.

Friend, M & Franson, JC (1999) *Field Manual of Wildlife Diseases - General Field Procedures and Diseases of Birds.* Library of Congress, Cataloging, USA.

King, JM (2014) The necropsy book A Guide for Veterinary Students, by The Necropsy Book. Charles Louis David DVM Foundation Publisher, Ithaca.

Munson, L (2000) *Necropsy of Wild Animals.* Wildlife Health Center, School of Veterinary Medicine, USA.

Somvanshi, R & Rao, JR (2009) *Necropsy techniques and necropsy conference manual.* Veterinary Research Institute, India.

Schmidt, RE & Reavill, DR (2003) *A Practitioner's Guide to Avian Necropsy.* Zoological Education Network, Lake Worth, Florida.

Work, TM (2000) *Manual de necropsia de tortugas marinas para biologos en refugios o areas remotas.* U. S. Geological Survey, National Wildlife Health Center Hawaii Field Station, Hawaii, USA.

White, CL & Dusek, RJ (2015) Wildlife Specimen Collection, Preservation, and Shipment. *Techniques and Methods* U.S. Geological Survey, U.S. Fish and Wildlife Service, and National Park Service, Virginia 32.
[http://dx.doi.org/10.3133/tm15C4]

CHAPTER 3

Cadaveric Phenomenons

Outline: In this chapter, we describe the main *post-mortem* alterations that can be observed in the process of decomposition of a cadaver.

Keywords: Animals Sentinels, Necropsy, *Post-Mortem*, Wildlife.

POST-MORTEM CHANGES

There are several difficulties in performing a necropsy in wild animals. Frequently the cadavers are not analysed because they were preyed upon, consumed by necrophages or died in a location that is difficult to access (*e.g.*, deep sea, tropical forest, deserts).

The greatest difficulty, however, is the decomposition of the corpse (Fig. **1**). The more time passes between the time of death and the time when the *post-mortem* exam is performer, the more putrefaction compromises the diagnosis (Woodford, Keet and Begins, 2000; Kinne, 2015; Garcês and Pires, 2017; Brooks, 2018).

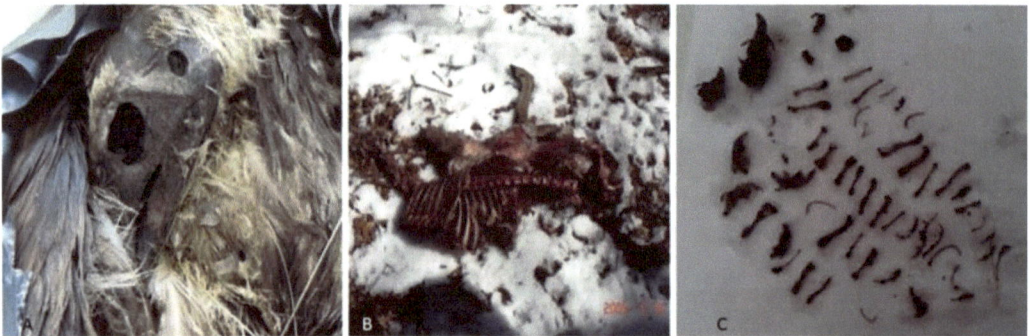

Fig. (1). Different examples of the advanced decomposition process, a mummified vulture (*Gyps fulvus*) **(A)**, road deer (*Capreolus capreolus*) skeleton **(B)** and small rodent bone remains in a pellet from a night bird prey **(C)**.

It is important to recognize the different stages of decomposition, not only to make an estimation about the time of death but also to understand if it is possible

Andreia Garcês & Isabel Pires

to collect viable samples to diagnose.

Some of the major factors that influence the process of decomposition are:

- **Temperature:** one of the most important factors in the progress of enzymatic digestion processes of the cadaver. Putrefaction is a biological process that works optimally in temperatures between 21°C to 38°C. Lower temperatures (under 10°C) and reduced ambient humidity limit enzymatic action, and bacterial and insect development. While it does not suppress decomposition, it manages to inhibit enzymatic activity enough to preserve the cadaver during a prolonged period. As such, the appearance of cadaveric phenomena is delayed. At temperatures below 0°C, the insects (adults and larva) die, although the larvae remain in body cavities such as skull, abdomen and vagina that can continue their activity as they can strengthen their metabolic health when present in large numbers. The body loses less biomass when it is in the shadow. High temperatures (summer), direct sunlight, combined with high ambient humidity significantly increase the degree of enzymatic action, bacterial and cadaverous flora and fauna. The inanimate body slowly loses temperature so that the *post-mortem* autolytic and post-rot processes occur rapidly.
- **Humidity:** also plays an important role in decomposition, particularly in mummification. The mummification of the body can be total or partial, and for its occurrence, the body must undergo an intense dehydration process, and have specific conditions, such as the existence of good ventilation and ideal temperatures to dry the tissues. Some locations (*e.g.*, deserts) have these ideal conditions. The rise of humidity directly relates to increased activity of insects (dipterous). So, humid and warm environments favour putrefaction while dry, cold or extremely hot environments slow the onset and disrupt it.
- **Animal Species:** the animal species not only determines the size of the individual, but also the type of external insulation, type of feed, degree of fat, *etc.* Equines and ruminants with their large body mass and the additional heat source of their digestive system lose temperature slowly, so the *post-mortem* changes initiate quickly. The elements that cover the body surface, such as skin attachments (hair, scales or feathers), usually prevent dissipation of body temperature and act as insulators. The abundant development of these appendages delays the loss of heat when compared with animals with a little cover of hair, wool or feathers. By delaying the elimination of body heat, both enzymatic and bacterial activities are more easily and quickly activated. For example, in canines with long hair, the corpse undergoes *post-mortem* changes faster than other canines of equal size but with short hair.
- **Body Size And Nutritional State:** in young individuals, heat loss happens faster than in adults. This is due to their higher rate of body surface area relative

to body mass. The higher the body surface area, the faster the cadaver cooling will be, while the greater the body, the slower the loss of body heat. Thus, in individuals of greater size or body mass, *post-mortem* changes occur and develop faster. This is mainly due to the longer time required to dissipate the internal heat, which is why the cadaver cooling will be slower. It is common to observe that in large animals, cadaveric phenomena develop much faster. In species with low body mass, cadavers are rapidly cooled, which means that *post-mortem* changes are delayed. Fatty or obese animals are characterized by having a thick layer of subcutaneous fat, this acts as a large insulating layer. The greater the degree of fat in the animal, the slower the loss of body heat from the corpse and thus the more rapid the appearance of cadaveric changes and putrefaction. In lean animals, comparatively, the processes start late because the heat loss of the corpse happens faster.

- **Health Condition:** certain diseases facilitate or favour the development of putrefaction. In subclinical disease states in which the animal dies rapidly and without clinical signs of disease (sudden death), or in those morbid conditions characterized by clinical signs of rapidly evolving disease, it is often observed that *post-mortem* changes start faster. This is because the animal usually finds the process of death with a very high content in the gastrointestinal tract or with high bacterial septicaemia. The opposite happens in those states of long-term illness or long agony; where the gastrointestinal content is usually low in quantity, as well as the fermentable or bacterial gastrointestinal load. The *post-mortem* changes are usually temporarily delayed in their appearance when compared to the previous case. Large wounds of traumatic origin favour the entry and subsequent dispersion of microorganisms (blood is a medium suitable for bacterial growth and dispersion) and deposition of eggs by cadaveric fauna. In diseases characterized by generalized oedema or tissue oedemas, the development of putrefaction is rapid, whereas in cases of severe dehydration, the onset of putrefaction is delayed.

- **Burial and Depth:** bodies deposited on the surface have a faster decomposition rate than the ones that are buried. The access of insects to bodies on the surface is faster. Bodies buried at a depth of 30 to 60cm become skeletons in a few months while bodies buried at 90 to120 cm take years to become only bone.

- **Scavenger Activity:** there are ghoulish, necrophiliac, omnivorous and opportunistic organisms or carrion fauna. They act both early and in the later periods of decomposition, collaborating in the destruction and elimination of the corpse. Some of the biological systems integrated with the corpse are:

- *Necrophages:* biological collectives that feed on corpses or dead organic matter (*e.g.*, vultures, crows).

- *Necrophiles:* predator biological organisms that parasitize or feed on the necrophagous community (*e.g.*, fungi).

- *Omnivores:* biological communities that feed on parts of the body occasionally, not depending on the body as food (*e.g.*, foxes, rodents, dogs).
- *Opportunists:* biological communities that use the corpse as a refuge.

These scavengers begin by eating the soft parts of the corpse's head, such as the eye and upper eyelids, ear, nose end, lips, jaw, tongue, *etc*. They also do so in the perineal region by taking the anus and the entire perianal muscle region, as well as destroying or eviscerating the organs contained in the pelvis. The skin and tissues clearly show the imprints of the teeth as well as the signs of tissues torn in *post-mortem* form (Huffman and Wallace, 2012; Cooper and Cooper, 2013; Nishant, Vrijesh and Ajay, 2018).

It is important to also refer the forensic entomology. The study of the invertebrate community in their different stages on the corpse is an important tool to determine the time of death. Each insect or group of insects will be attracted to the process during one phase or another of the cadaverous decomposition, mainly based on their nutritional needs. The species will change with the geographical location and season.

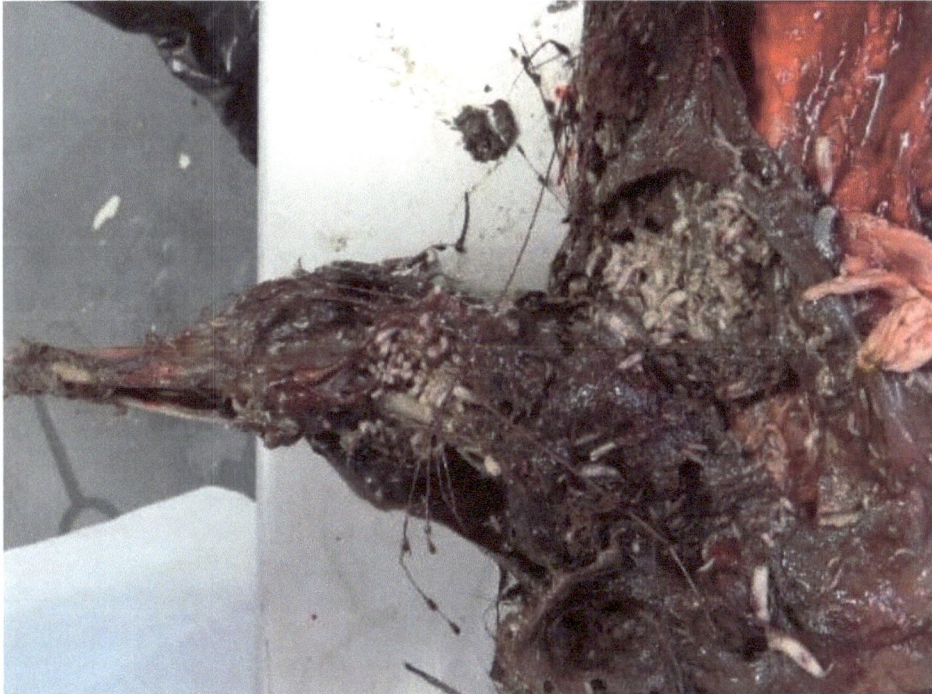

Fig. (2). Diptera larva in a cadaver.

The main arthropods of forensic interest are the following:

DIPTERA (CLASS INSECTA): they have two membranous anterior wings and two posterior wings reduced (halters). It is the group of flies and mosquitoes (rarely, they intervene in the cadaveric decomposition). They are the insects that have the greatest forensic interest, as well as the first to arrive at the corpse. They mostly feed on dead tissues (Fig. **2**).

COLEOPTERA (CLASS INSECTA): they have the first pair of hardened wings (elytra), forming a kind of carapace (Fig. **3**). Among the Coleoptera of forensic interest, some feed on cadaveric remains, for example, the genus *Dermestes*.

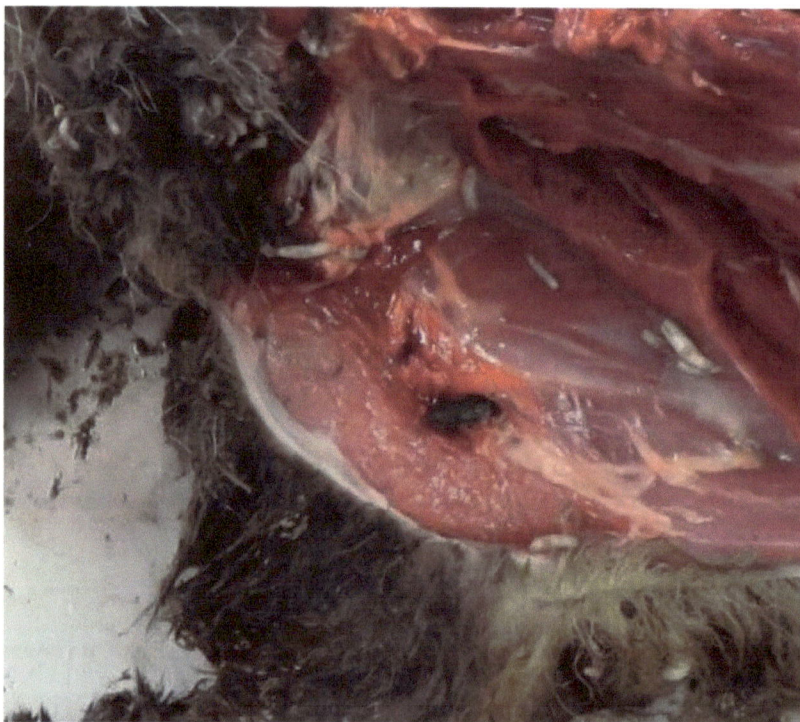

Fig. (3). Adult Cloptera in a corpse.

HYMENOPTERA (CLASS INSECTA): they have membranous wings This includes several groups, with ants and wasps being the ones most often found in corpses. They feed on most of the larvae of other insects.

MITES (CLASS ARACHNIDA): these are mainly present on the earth under the corpse, feeding on remains of dried tissues and other insects.

Other groups of insects can be present, but they usually less used in the determination of the time of decomposition (Zachary, 2016; Garcês and Pires, 2017; McDonough and Southard, 2017; Brooks, 2018).

The processes of biodegradation of body organic matter or cadaveric transformation start immediately after death and continue to occur over a prolonged period at different rates for different organs. These processes are called cadaveric alterations or *post-mortem* changes. Cadaverous Phenomena can be classified as abiotic when there is no intervention by microbiological agents, and biotic agents, where microbiological agents will cause changes in profound aspects of the corpse's appearance and structure. Abiotic phenomena are further classified as immediate and intermediate or consecutive. *Post-mortem* changes are classified based on their order of appearance in (Table **1**).

Table 1. Classification of cadaveric phenomenons, by chronology of its presentation.

Abiotic Cadaveric Phenomena		Biotic Cadaveric Phenomena	
Immediate Phenomena	**Mediate or Consecutive Phenomena**	**Transformative**	**Conservative**
Stopping cardiorespiratory function Brain death Insensitivity (tactile, thermal, painful), Immobility, Absence of breathing and blood circulation, Loss of muscle tone (with mydriasis and sphincter relaxation)	Cadaverous lividness or *Palor mortis* Cadaverous cooling, *Algor mortis* Cadaverous stiffness or *Rigor mortis* Cadaverous evaporation Blood coagulation Hypostasis (*Livor mortis*) and cadaverous spots	Chromatic or putrefaction spots Gaseous, emphysematous or deforming development Putty or putrid fusion Skeletal reduction	Maceration Mummification Saponization

The *immediate abiotic phenomena* happen after apparent death, such as loss of consciousness, insensitivity, and immobility, loss of muscular tonus, breathe and blood circulation suppression.

THE ABIOTIC MEDIATE OR CONSECUTIVE PHENOMENA

Follow the immediate ones and correspond to all those morphological or structural changes that occur early in the body, starting at the moment of death and continuing to appear until the first signs of cadaveric deterioration. Their evolution depends on several factors. They are:

• **Cadaveric Dehydration:** after death, skin and mucous membranes dry. These lose their natural moisture by evaporation, a phenomenon that is directly related to environmental factors (*e.g.*, temperature, humidity and ventilation). The skin hardens, losing its natural elasticity and taking a hard (leathery) consistency. Upon observation, the mucous membranes (ocular, buccal and genital) appear dry, opaque and tend to obscure from their normal colouration. The eyeballs retract (ocular sinking) and the cornea dries out, losing its natural transparency (corneal clouding). This phenomenon is dependent on ambient temperature and humidity.

- *Algor Mortis:* after death occurs, the body temperature decreases until it reaches the environmental temperature. This cadaveric cooling is a direct consequence of the disappearance of the natural mechanisms of generation and regulation of body temperature. Together with the cessation of blood circulation, muscle activity and changes in tissue metabolic activity, these lead to accentuated cadaveric cooling. The ambient temperature is a critical factor that affects the rate of *post mortem* cooling of the body. Cooling will be faster in a body immersed in water or thin bodies and slower in obese bodies.

- **Rigor Mortis:** is the stiffening of the muscles of the corpse due to depletion of adenosine triphosphate (ATP) after death and the subsequent build-up of lactate, which results in an inability to release the actin-myosin bond. This process involves all muscles of the body; however, it generally becomes apparent first in smaller muscles such as the jaw and the eyelids due to a faster depletion of ATP comparatively to the larger muscles of the trunk and limbs. Cadaveric stiffness begins approximately 2-3 hours after death, but there is a large margin of variation, regarding both the time of onset and its average duration. Taking into account the different phases, from beginning to end, the cadaverous stiffness can occur between 18-24 to 36-48 hours. The duration of this process can be influenced by many factors, such as ambient temperature and the degree of muscular activity before death.

- *Livor Mortis (Post Mortem* **Hypostasis):** this phenomenon is characterized by the appearance of vinous or bluish-red spots on the skin of the lower areas of the body. This is due to the gravitational sedimentation of blood with intravascular pooling, especially in the capillary bed and veins. This is a passive event, which develops as a direct consequence of the loss of vascular tone and the effect of gravity, with sedimentation and stagnation of blood still flowing. They become noticeable between 30 minutes and 4 hours. post-death; and manifest intensely between 8 and 12 hours after death.

- **Visceral Hypostasis:** this is characterized by gravitational sedimentation of non-circulating intravascular blood, with subsequent accumulation in the visceral areas of decline. It is very commonly observed in the lungs, liver, kidneys and intestines.

- *Post-mortem* **Autolysis:** a process of dissolving tissues and cells quickly begins with death, whether they are normal or diseased and regardless of previous injuries. This biological phenomenon is characterized by the fact that the dissolution or digestion of tissues and cells is affected by their cellular enzyme systems.

- **Destruction of the Corpse by Exogenous Factors:** the corpse is usually destroyed or altered by the action of certain exogenous factors during the early stages of cadaveric evolution - climatic or environmental conditions, fauna and flora (Zachary, 2016; Garcês and Pires, 2017; McDonough and Southard, 2017; Brooks, 2018).

BIOTIC CADAVERIC PHENOMENA

This phenomena is associated with autolytic phenomena and intense microbial proliferation, mainly from the digestive tract of cadavers, from sources of infection in this or the environment, and are characterized by intense changes in the shape, colour and structure of the corpse, culminating in with the decomposition of organic matter and reduction of the corpse to the skeleton.

- *Post-Mortem* **Impregnation:** is the *post-mortem* colouration that some tissues take by degraded organic pigments, specifically haemoglobin, which is released during the autolysis and cellular digestion processes with subsequent rot. This post-mortem change appears, specifically, between the middle or advanced period of the post-mortem autolysis and the first stage of putrefaction.
- **Pseudomelanosis:** is the appearance of a greyish, greenish or blackish colour in the tissues after an advanced rot state. The production hydrogen sulphide by bacteria combined with the iron in haemoglobin produces iron sulphide, a black pigment that imparts this blackish brown colour to the tissues.
- *Post Mortem* **Emphysema:** the accumulation of gas in the cavities, the hollow tissues and solid organs of the body. The gas comes from bacterial fermentation and rot. The gases produced during the rotting process are methane, carbon dioxide, ammonia, sulfuric gas, *etc*. The smell of rot is produced by these gases and small amounts of mercaptan.
- *Post-Mortem* **Rupture:** the fermentation gases that are produced and accumulated in the hollow cavities cause the progressive distension of the body structures. This eventually leads to the rupture of organs such as the stomach, intestines, *etc*., due to the extreme pressure caused by the accumulation of gases and also by the weakening of the walls as a result of autolysis.
- *Post-Mortem* **Displacement:** both the accumulation of gases in the hollow visors and the movements (moved, rotated or hung) of the corpse occasionally cause displacement of viscera; especially in the intestinal mass.
- **Putrefaction:** it is a complex biological process that is characterized especially by the *post-mortem* decomposition or degradation of the tissues and all the

organic matter component of a corpse. Tissue destruction and organic matter are the result of the combined action of endogenous enzymes (autolysis) and a potent enzymatic and degradative action exerted by an important flora and fauna composed of bacteria, fungi, arthropods and vertebrates. All this flora and fauna act on the basic components of the dead animal organism and body fluids until they achieve their total degradation (Zachary, 2016; Garcês and Pires, 2017; McDonough and Southard, 2017; Brooks, 2018).

Necropsy Techniques for Examining Wildlife Samples, 2020, x-xx

Fresh phase	Bloated phase	Active decay phase	Advanced decay phase	Skeletal reduction phase
1	2	3	4	5

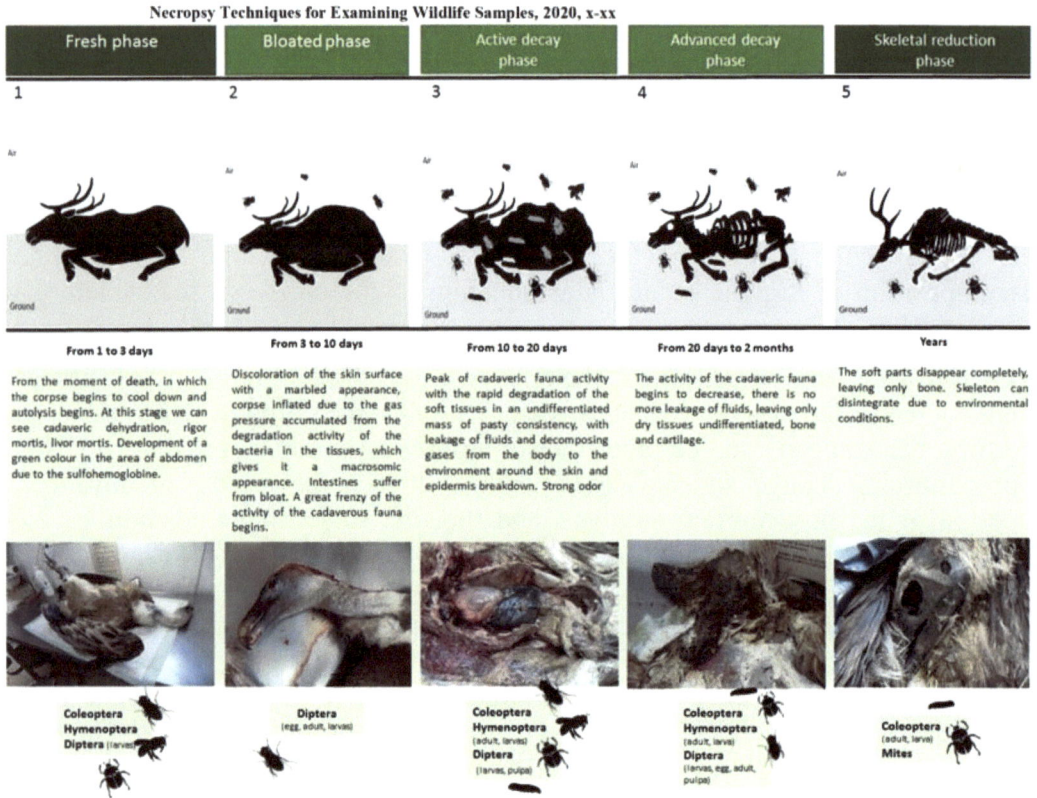

From 1 to 3 days	From 3 to 10 days	From 10 to 20 days	From 20 days to 2 months	Years
From the moment of death, in which the corpse begins to cool down and autolysis begins. At this stage we can see cadaveric dehydration, rigor mortis, livor mortis. Development of a green colour in the area of abdomen due to the sulfohemoglobine.	Discoloration of the skin surface with a marbled appearance, corpse inflated due to the gas pressure accumulated from the degradation activity of the bacteria in the tissues, which gives it a macrosomic appearance. Intestines suffer from bloat. A great frenzy of the activity of the cadaverous fauna begins.	Peak of cadaveric fauna activity with the rapid degradation of the soft tissues in an undifferentiated mass of pasty consistency, with leakage of fluids and decomposing gases from the body to the environment around the skin and epidermis breakdown. Strong odor	The activity of the cadaveric fauna begins to decrease, there is no more leakage of fluids, leaving only dry tissues undifferentiated, bone and cartilage.	The soft parts disappear completely, leaving only bone. Skeleton can disintegrate due to environmental conditions.
Coleoptera Hymenoptera Diptera (larvae)	Diptera (egg, adult, larvas)	Coleoptera Hymenoptera (adult, larvas) Diptera (larvas, pupla)	Coleoptera Hymenoptera (adult, larva) Diptera (larvas, egg, adult, pupla)	Coleoptera (adult, larva) Mites

Fig. (4). Different phases that experience a corpse from the moment of death until it degrades completely (the time in the scheme is related to the composition of a carnivore animal of medium size, this period will change as previously described). Description of the phases and cadaveric fauna involved.

(Fig. 4) is a simplified summary of the different phases that experience a corpse from the moment of death until it degrades completely (the time in the scheme is related to the composition of a medium-size carnivore, this period will change as depending on various factors as previously described).

The corpses that are submerged in water have a rhythm of decomposition different from those that are exposed to the air due to several factors in addition to those already indicated previously. The process of decomposition in water is delayed since the body temperature is lost twice as fast as when exposed to air. Unlike what occurs in the earthly environment, the process begins in the head as it is tilted downwards (blood accumulates in this zone), then goes to the neck area. The black-green colouration of the tissues only appears 6-10 days after death.

The cadaverous fauna in these cases is a little different than what happens when exposed to air. Since it only has access to the body when it is floating and there is no equivalent sarcosaprophagous fauna in the water, there are only a few species that can act as cadaveric fauna. Their proliferation in the body will depend on many factors such as water flow, substrate characteristics and the position of organisms in the water column (*e.g.*, trichoptera).

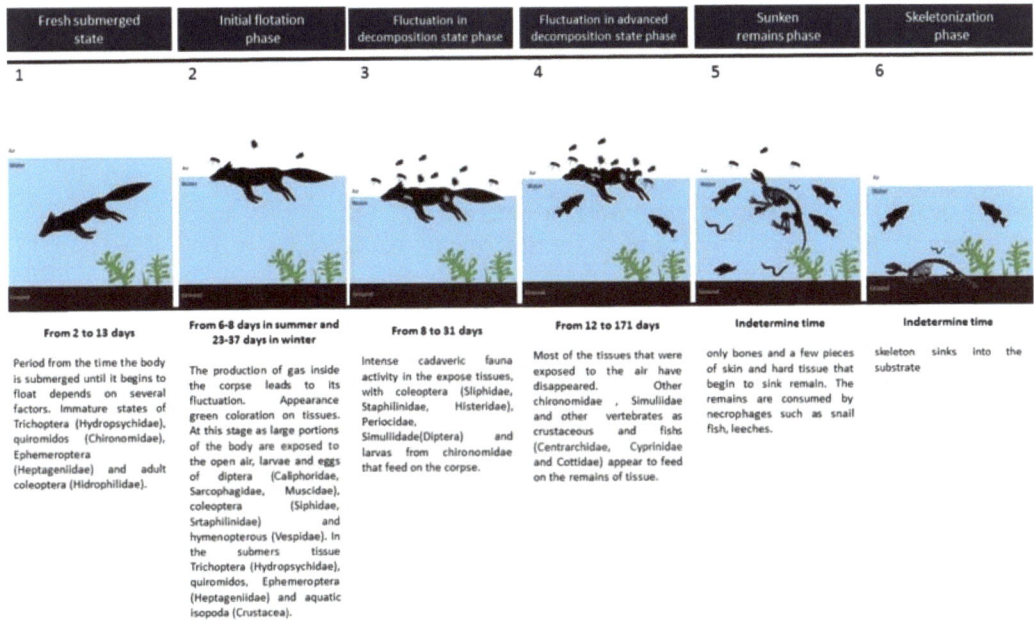

Fresh submerged state	Initial flotation phase	Fluctuation in decomposition state phase	Fluctuation in advanced decomposition state phase	Sunken remains phase	Skeletonization phase
1	2	3	4	5	6

From 2 to 13 days	From 6-8 days in summer and 23-37 days in winter	From 8 to 31 days	From 12 to 171 days	Indetermine time	Indetermine time
Period from the time the body is submerged until it begins to float depends on several factors. Immature states of Trichoptera (Hydropsychidae), quiromidos (Chironomidae), Ephemeroptera (Heptageniidae) and adult coleoptera (Hidrophilidae).	The production of gas inside the corpse leads to its fluctuation. Appearance green coloration on tissues. At this stage as large portions of the body are exposed to the open air, larvae and eggs of diptera (Caliphoridae, Sarcophagidae, Muscidae), coleoptera (Siphidae, Srtaphilinidae) and hymenopterous (Vespidae). In the submers tissue Trichoptera (Hydropsychidae), quiromidos, Ephemeroptera (Heptageniidae) and aquatic isopoda (Crustacea).	Intense cadaveric fauna activity in the expose tissues, with coleoptera (Sliphidae, Staphilinidae, Histeridae), Periocidae. Simullidade(Diptera) and larvas from chironomidae that feed on the corpse.	Most of the tissues that were exposed to the air have disappeared. Other chironomidae , Simuliidae and other vertebrates as crustaceous and fishs (Centrarchidae, Cyprinidae and Cottidae) appear to feed on the remains of tissue.	only bones and a few pieces of skin and hard tissue that begin to sink remain. The remains are consumed by necrophages such as snail fish, leeches.	skeleton sinks into the substrate

Fig. (5). Different phases that experience a corpse from the moment of death until it degrades completely in the aquatic environment (the time in the scheme is related to the composition of a carnivore animal of medium size, this period will change as previously described). Description of the phases and cadaveric fauna involved.

In the marine water environment, the cadaverous fauna is mainly constituted by crustaceans, fish, gastropod molluscs and echinoderms. Factors affecting decomposition are water salinity, depth, currents, substrate nature and several scavengers (Zachary, 2016; Garcês and Pires, 2017; McDonough and Southard, 2017; Brooks, 2018).

In aquatic environments, 6 phases of decomposition are considered (the time is relative, it changes according to the factors described previously) (Fig. **5**).

REFERENCES

Brooks, JW (2018) *Veterinary Forensic Pathology.* Springer International Publishing, NY, USA.

Cooper, J & Cooper, ME (2013) *Wildlife forensic investigation: Principles and Practice.* CRC Press Taylor & Francis Group, Boca Raton.
[http://dx.doi.org/10.1201/b14553]

Garcês, A & Pires, I (2017) Manual de Técnicas de Necrópsia em *Animais Selvagens.* Arteology, Porto.

Huffman, J & Wallace, J (2012) *Wildlife forensics: Methods and Applications.* Wiley-Blackwell, West Sussex.

Kinne, J (2015) Post mortem Examination.*Avian Medicine* Mosby, USA, 567-81.

McDonough, SP & Southard, T (2017) *Necropsy Guide for Dogs, Cats, and Small Mammals.* Wiley-Blackwell, Ames, Iowa.
[http://dx.doi.org/10.1002/9781119317005]

Nishant, K, Vrijesh, K & Ajay, K (2018) Wildlife Forensic: Current Techniques and their limitations. *Journal of Forensic Science & Criminology,* 5.
[http://dx.doi.org/10.15744/2348-9804.5.402]

Woodford, MH, Keet, DF & Begins, RG (2000) *Post-mortem procedures for wildlife veterinarians and field biologists,* Iucn, Paris, France.

Zachary, JF (2016) *Pathologic basis of veterinary disease.* Elsevier, USA.

CHAPTER 4

Necropsy in Wild Birds

Outline: In this chapter, we describe the method of necropsy in wild birds, offering some information regarding the different orders and anatomic characteristics of determining species.

Keywords: Aquatic Birds, Animals Sentinels, Conservation, Mortality, Necropsy, Pathology, *Post-Mortem*, Wild Birds.

GENERAL CONSIDERATIONS

In wild and domestic birds, necropsy should be performed as early as possible since putrefaction occurs rapidly after death due to the high a*nte-mortem* body temperature (40 °C), their fast metabolism and the thermal insulation effect of the feathers (Munson, 2000; Garcês and Pires, 2017). If it is not possible to perform the procedure at the moment, the carcass must be refrigerated. The carcass should be immersed in cold water to eliminate the isolation capacity of the feathers before refrigeration (Fig. **1**) (King, *et al.* 2014). Freezing the carcass is not recommended because, during the histological analysis, the ice crystals destroy the structure of the cell.

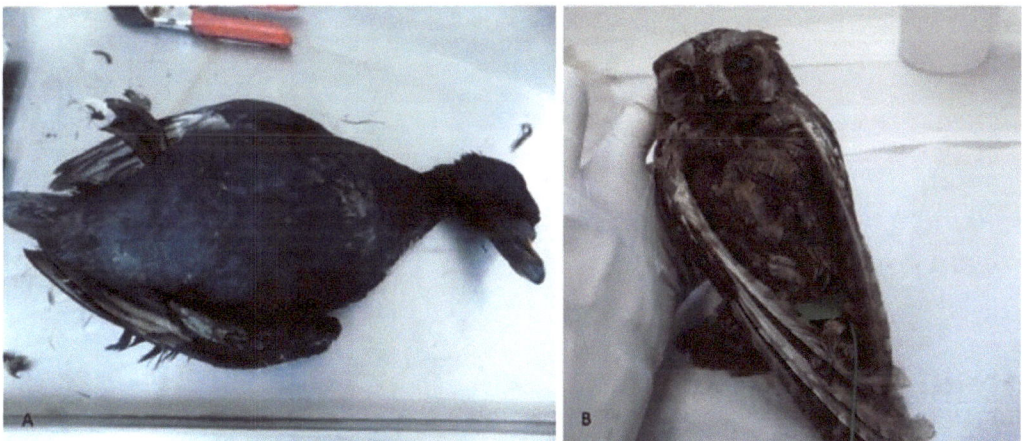

Fig. (1). In **A**, a *Melanitta negris* frozen cadaver and in **B** a *Otus scops* non frozen cadaver.

Andreia Garcês & Isabel Pires

However, in the case of freezing, some data and material can still be collected, for instance, material for toxicological analyses (King, *et al.* 2013; Brooks, 2018).

EXTERNAL EXAM

Before the procedure is started, it is beneficial to perform an X-ray, especially if there is a suspicion of a traumatic death involving a firearm or a metal object such as a hook. Besides, other lesions may be visible as fractures or granulomas (Friend and Franson, 1999; Schmidt and Reavill, 2003; Dorrestein, 2008; Butcher and Miles, 2015).

In the external exam, the species must be identified and its conservation status should be known as it may be necessary to alert the responsible authorities. It is also important to know the behaviour, diet and natural habitat of the animal (Harcourt-Brown and Chitty, 2005; Chitty and Lierz, 2008; Chitty and Monks, 2018).

Beyond animal weight, the body condition should be evaluated based on the palpation of the pectoral muscles and the exposure of the keel (Fig. **2**) (Harcourt-Brown and Chitty, 2005; Chitty and Lierz, 2008; Chitty and Monks, 2018).

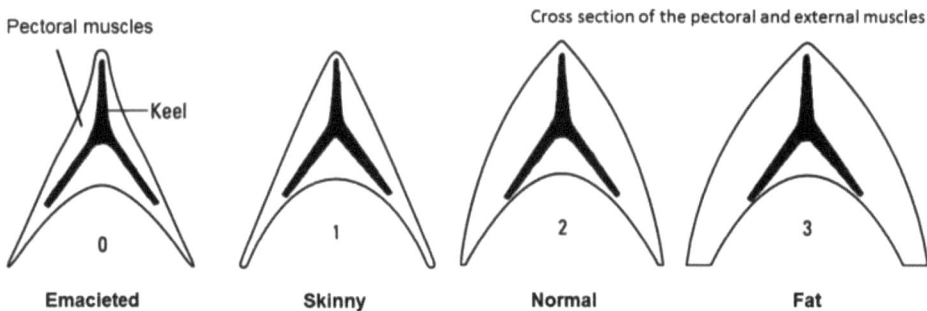

Fig. (2). Scheme of the body condition in birds, based in the appearance of the pectoral muscles.

In birds, it is possible to determine whether they are adults, subadults, juveniles or pullets through their plumage, therefore allowing for an estimate of age. In some species, it is even possible to determine whether they are males or females based on their phenotypic characteristics (Figs. **3** and **4**) (Friend and Franson, 1999; Schmidt and Reavill, 2003; Dorrestein, 2008; Butcher and Miles, 2015).

If there is no previous information (biometrics, genetics or endoscopy), it is simpler to determine the gender of a bird during the internal exam. Some species present incubation plates, which can be observed above the cloaca as an area of skin without feathers, thickened and very vascularized (*e.g. Bubo bubo*) (Fig. **5**).

These can exist only in females or only in males, depending on the species (Harcourt-Brown and Chitty, 2005; Chitty and Lierz, 2008; Chitty and Monks, 2018).

After identification, the animal should be weighed, measured (biometrical measurements) and evaluated externally by visual examination and palpation (Harcourt-Brown and Chitty, 2005; Chitty and Lierz, 2008; Chitty and Monks, 2018).

Fig. (3). Representation of the beak in species of birds of different orders, adapted to different diets: **A**-*Asio otus*, **B**-*Alcedo atthis,***C**-*Phoenicopterus roseus* and **D**-*Fratercula arctica*.

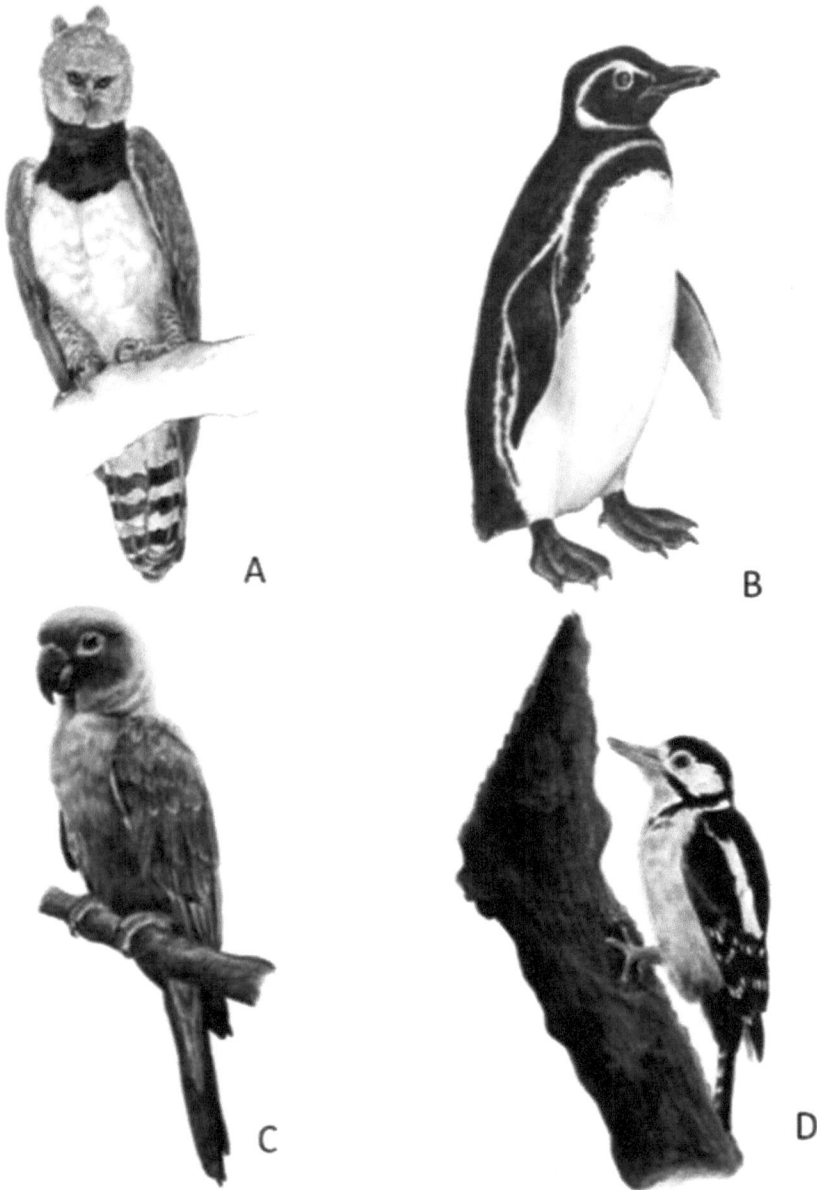

©Andreia Garcês

Fig. (4). Representation of species of birds of different orders: **A**-*Harpya harpyja*, **B**-*Spheniscus magellanicus*, **C**-*Arantinga solestiallis* and **D**-*Dendrocopos major.*

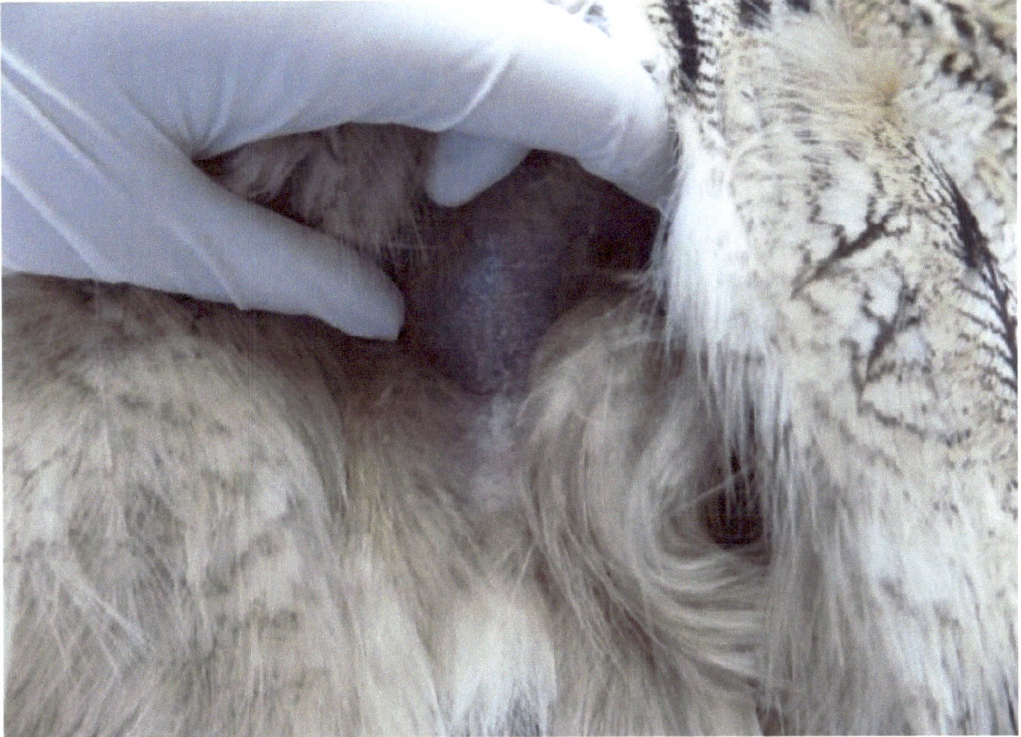

Fig. (5). Incubation plates in a *Bubo bubo.*

ᴛ All aspects of external habit should be assessed, such as the presence of identifying marks such as rings or tattoos, the feather condition, skin, claws, natural orifices, eyes and joints (Fig. **6**).

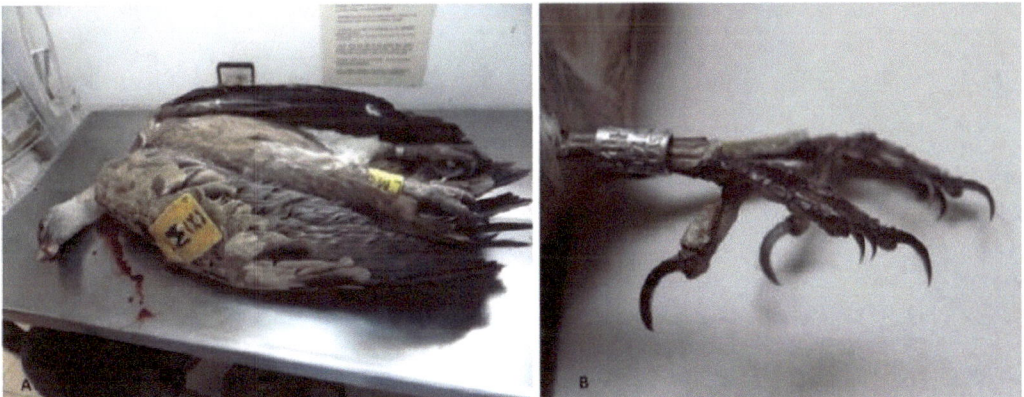

Fig. (6). Identification marks on a necrophagous bird (**A**) and a passerine (**B**).

The plumage condition, the existence of broken feathers, or if they are dirty with faeces, burnt, gnawed or covered by substances such as water, blood, oil, etc, among other changes should all be noted. In the case of unknown substances, some feathers should be collected for later identification (Figs. **7** & **8**) (Friend and Franson, 1999; Schmidt and Reavill, 2003; Dorrestein, 2008; Butcher and Miles, 2015).

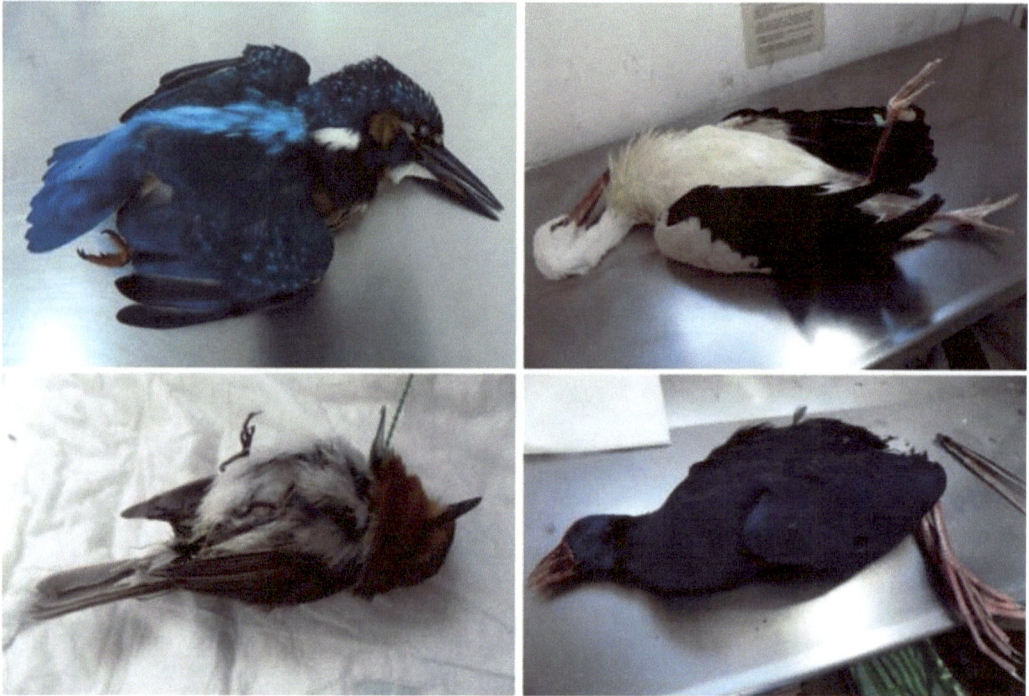

Fig. (7). Examples of the external habit of birds of different orders.

If ectoparasites are found, they should be collected and later identified. The palpation of the carcass could reveal dislocations, fractures, amputations, wounds, foreign bodies, emphysema, subcutaneous masses or some deformity. It should be noted that the skin of birds is thinner and fragile than other vertebrates such as mammals (Harcourt-Brown and Chitty, 2005; Chitty and Lierz, 2008; Chitty and Monks, 2018).

Fig. (8). Detailed external examination of the eye, ear, hind limb and oral cavity of a nocturnal bird of prey.

All-natural orifices should be inspected systematically: eyes, ears, nostrils, oral cavity, cloaca, uropygial gland (Friend and Franson, 1999; Schmidt and Reavill, 2003; Dorrestein, 2008; Butcher and Miles, 2015). The wax present in the dorsal base of the maxillary mandible is keratin. This structure may have different shapes and colours depending on the species and sex. Also, the lesions presented in this wax can indicate fights with other animals or captivity (Harcourt-Brown and Chitty, 2005; Chitty and Lierz, 2008; Chitty and Monks, 2018).

The uropygial gland (present at the base of the tail, on the last vertebra) is not present in the orders Struthioniformes, Rheidaformes, Casuariiformes and Otididae. It is very well developed in Sphenisciformes, Podicipediformes and

Laridae. It may be small or absent in some animals belonging to the orders Caprimulgiformes, Columbiformes, Psittaciformes and Piciformes. In marine birds, we can also observe the presence of the very developed salt gland, located above the eyes (Fig. **9**) (Friend and Franson, 1999; Schmidt and Reavill, 2003; Dorrestein, 2008; Butcher and Miles, 2015). Figs. (**10-14**) showed the external examination of different bird species.

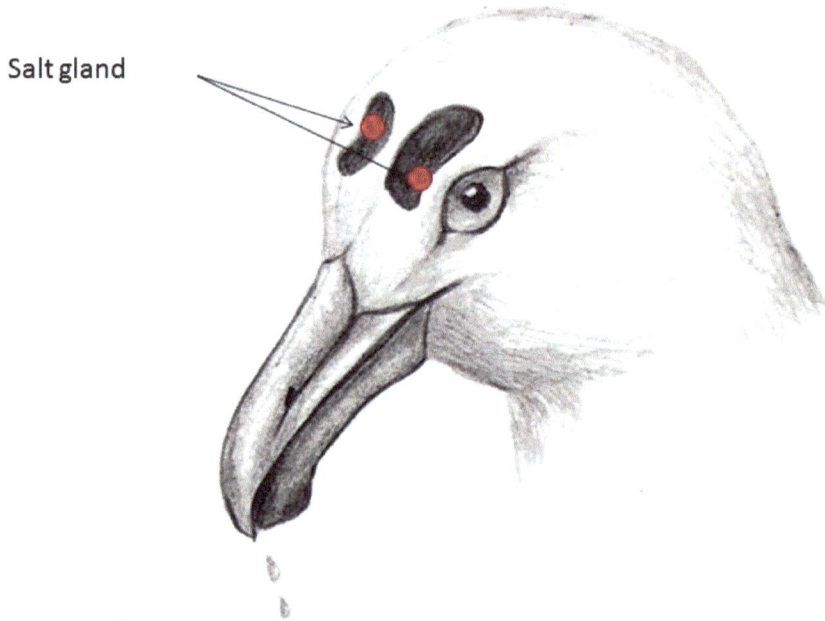

Salt gland

Fig. (9). Representation of the salt gland in seabirds.

A

Fig. 10 cont.....

© Andreia Garcês

Fig. (10). Representation of external examination of a bird carcass (*Alcedo atthis*), in ventrodorsal position (**A**) and dorsoventral position (**B**).

Skin samples (3-5 mm thick when possible) can be taken to place in 10% formaldehyde. At this stage, whenever possible, scraping and swabs of the cloacal and oral cavity should be performed for bacterial and parasite observation. In the case recently dead animals, an ophthalmic examination can be performed (Figs. **10-14**) (Friend and Franson, 1999; Schmidt and Reavill, 2003; Dorrestein, 2008; Butcher and Miles, 2015).

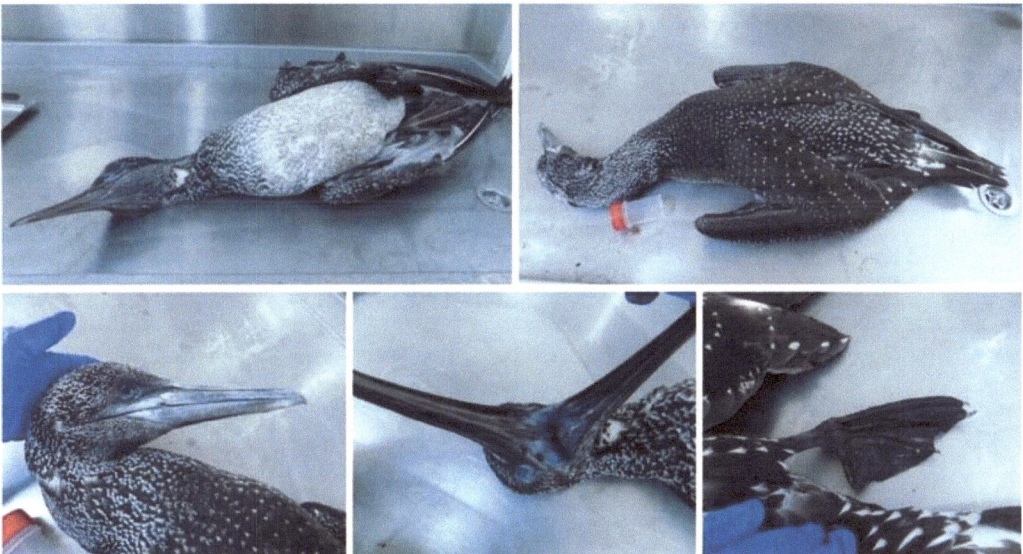

Fig. (11). External exam in a *Morus bassanus*.

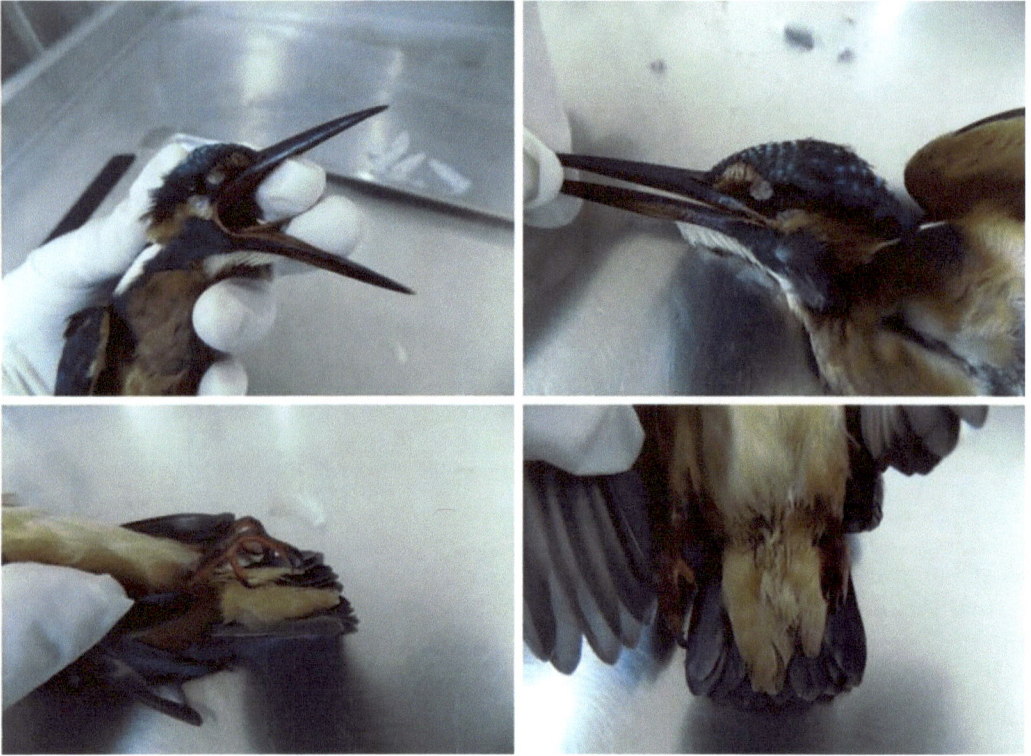

Fig. (12). External exam in an *Alcedo atthis.*

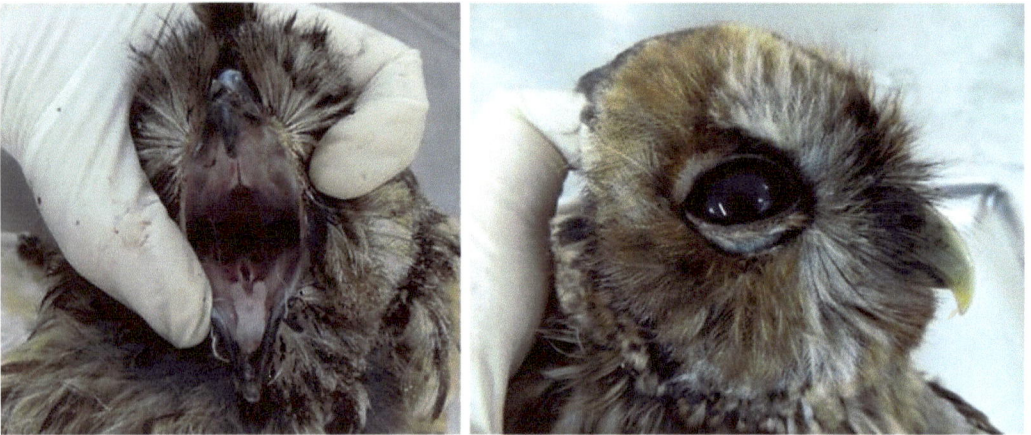

Fig. 13 cont.....

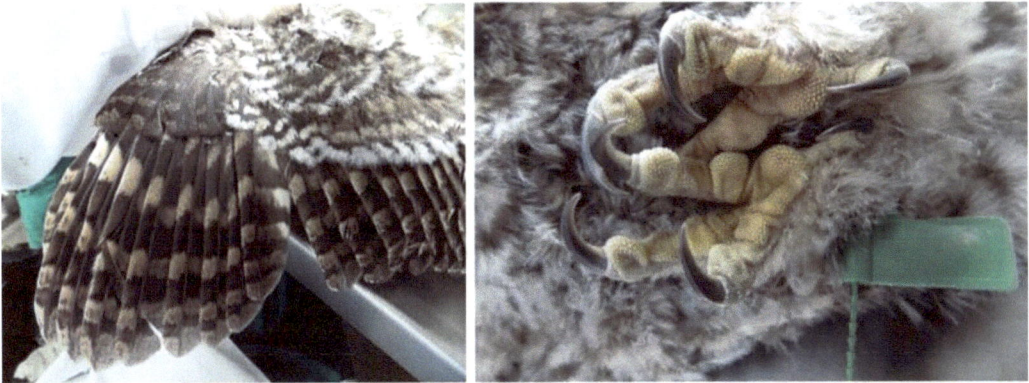

Fig. (13). External exam in a *Strix aluco*.

Fig. (14). External exam in an *Ara chloropterus*.

INTERNAL EXAM

To decrease feather debris during handling, which may contain *Chlamydophila* spp. or other organisms that can contaminate the viscera, the feathers should be wet or the carcass immersed in 70% alcohol. (Friend and Franson, 1999; Schmidt and Reavill, 2003; Dorrestein, 2008; Butcher and Miles, 2015).

The animal should be placed in dorsoventral recumbency. The midline feathers should be removed to make the incision easier, without damaging the viscera and making it possible to identify subcutaneous lesions such as bruising and lacerations. In ducks, cormorants or geese, because they have a very dense plumage and the skin easily detaches from the muscle, it is easier to make the incision directly on the skin similar to what is performed in mammals (Harcourt-Brown and Chitty, 2005; Chitty and Lierz, 2008; Chitty and Monks, 2018).

In larger species, the coxofemoral joints are disarticulated through an incision in the skin, adductor muscle and capsule of the coxofemoral joint. The knees are then forced craniolaterally so that the carcass is stable (Fig. **15**) (Friend and Franson, 1999; Schmidt and Reavill, 2003; Dorrestein, 2008; Butcher and Miles, 2015).

Fig. (15). Schematic of the incision for exposure of the pectoral muscles, with these already evidenced (*Alcedo atthis*).

In smaller passerine species, it may be easier to attach the wings and back limbs to a plate with needles or pins (Harcourt-Brown and Chitty, 2005; Chitty and Lierz, 2008; Chitty and Monks, 2018).

An incision is made, from the intermandibular area to the pelvic area, next to the cloaca. The skin should be removed through the blunt dissection to expose the cervical, pectoral, abdominal muscles, trachea, oesophagus, craw and keel (Friend and Franson, 1999; Schmidt and Reavill, 2003; Dorrestein, 2008; Butcher and Miles, 2015).

The muscles, especially the pectoral muscles, should be observed to assess the presence of lesions. The normal colouration of the skeletal muscles is reddish-brown. Longitudinal incisions should be made to observe possible deep lesions, such as hematomas, haemorrhages, parasites, projectiles or granulomas (Harcourt-Brown and Chitty, 2005; Chitty and Lierz, 2008; Chitty and Monks, 2018) (Figs. **15-17**).

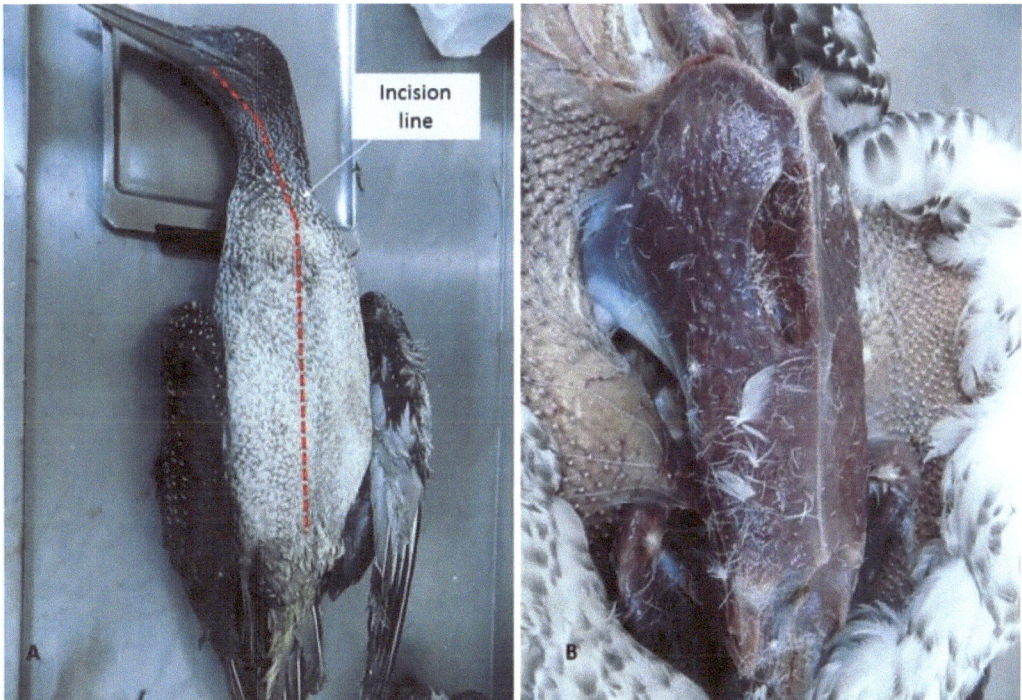

Fig. (16). Representation of the incision line in a *Morus bassanus.*

Skeletal muscles should be observed; their normal colouration should be pink. If these are red, hyperaemia may be indicative of septicaemia; if they are pale, it may indicate anaemia. Dry muscles occur in dehydration and if they appear irregular and atrophied, it may indicate cachexia. If there is unilateral discolouration of the muscles by hypostasis, it may indicate that the animal has been lying on that side for some time after death (Harcourt-Brown and Chitty, 2005; Chitty and Lierz, 2008; Chitty and Monks, 2018).

To access the coelomic cavity, a longitudinal incision is made at the level of the coracoid bone, along with the pectoralis muscle, on each side of the thorax. With forceps, the abdominal muscle should be elevated slightly and with a scalpel a transverse incision is made near the caudal zone of the sternum, being careful not to lacerate the liver (Harcourt-Brown and Chitty, 2005; Chitty and Lierz, 2008; Chitty and Monks, 2018). Then the pectoral and keel muscles are removed in a block cutting the ribs, coracoid bones and clavicle at the junction of these with the sternum, with the help of scissors or pliers depending on the size of the animal. In fresh cadavers, care should be taken not to cut the brachiocephalic arteries, which would cause blood to enter the lungs through the thoracic air sacs (Friend and Franson, 1999; Schmidt and Reavill, 2003; Dorrestein, 2008; Butcher and Miles, 2015).

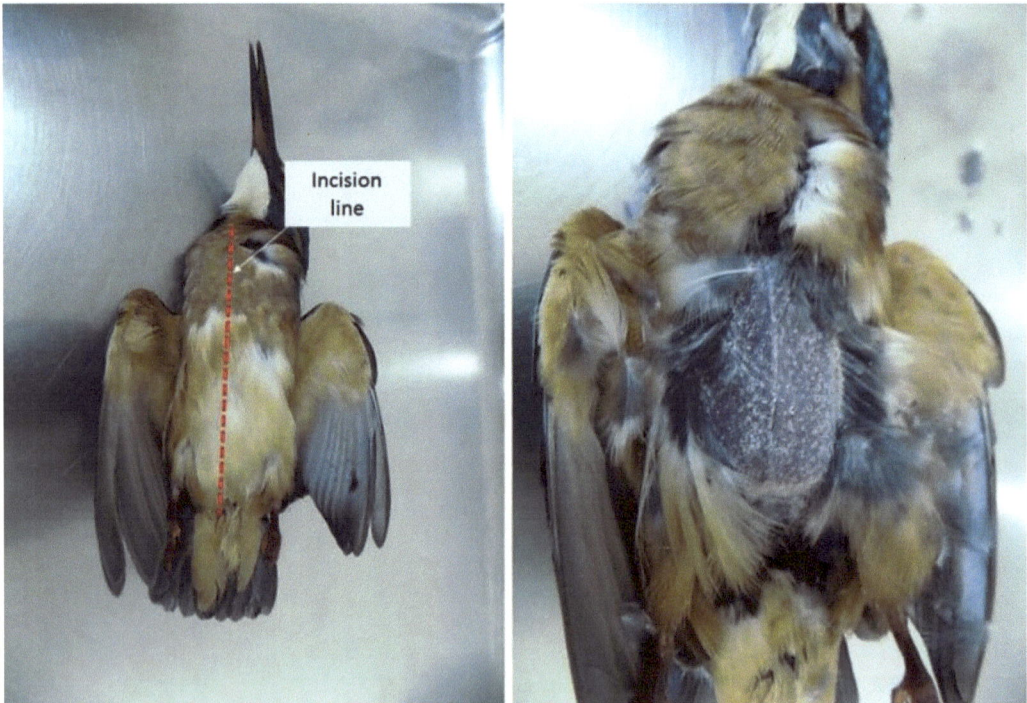

Fig. (17). Representation of the incision line in an *Alcedo atthis*.

Once the sternum is raised, it is possible to observe the thoracic and abdominal air sacs. Their normal appearance is of a moist and transparent membrane. If they are abnormal or opaque, samples should be collected for microbiological or cytological examination (Friend and Franson, 1999; Schmidt and Reavill, 2003; Dorrestein, 2008; Butcher and Miles, 2015). If the material is collected for histological examination, it should be placed on a sheet of paper before fixing, this allows its identification in the processing and minimizes the possibility of losing these membranes (Harcourt-Brown and Chitty, 2005; Chitty and Lierz, 2008; Chitty and Monks, 2018).

After the procedure described, the internal organs are exposed (Figs. **18 & 19**). These should be examined visually *in situ*. The presence of fluids or exudates within the coelomic cavity should be evaluated. The presence of a minimal amount of clear fluid is normal. The presence of a small amount of fat in the abdomen is also normal. Excess fat indicates that it is an obese animal (probably animal kept in captivity, with too much or poor nutritional value food). In cachexia, serous atrophy of the fat is detected. Peritonitis caused by abdominal posture can be detected by the presence of large amounts of yellow material on the intestinal serosa. The presence of urate crystals on the organs may be indicative of visceral gout (Friend and Franson, 1999; Schmidt and Reavill, 2003; Dorrestein, 2008; Butcher and Miles, 2015).

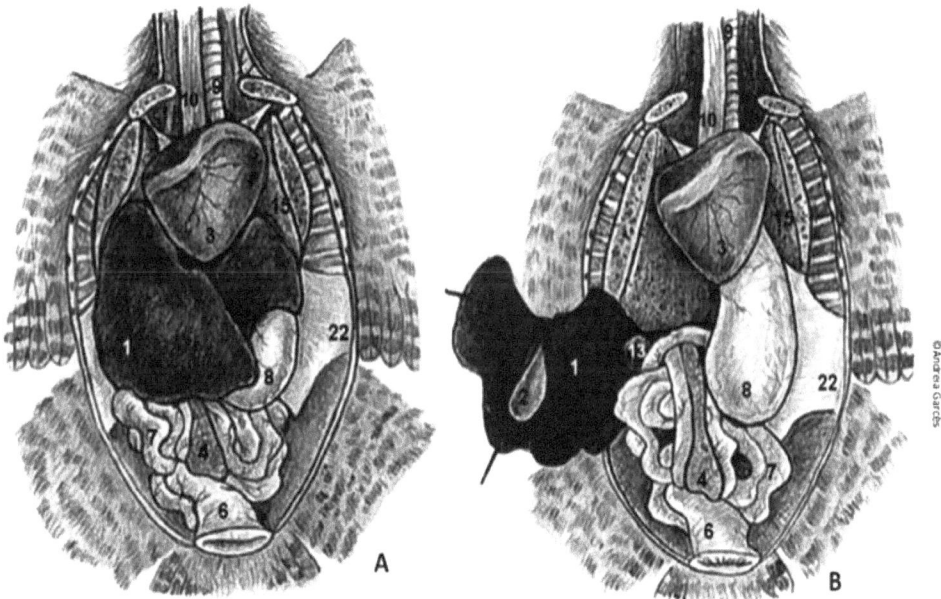

Fig. (18). Schematic representation of the organs *in situ* after removal of the keel, (1) - liver; 2 - biliary vesicle; 3 - heart; 4 - pancreas; 5 - kidney; 6 - large intestine; 7 - small intestine; 8 - stomach; 9 - trachea; 10 - oesophagus; 11 - thyroid; 12 - parathyroid; 13 - spleen; 15 - lung; 16 - testicles; 17 - adrenal gland; 20 - ovary; 21 - oviduct, 22 - air-sacs.

After observing the viscera in *locus,* they should be removed, one by one for a more detailed examination.

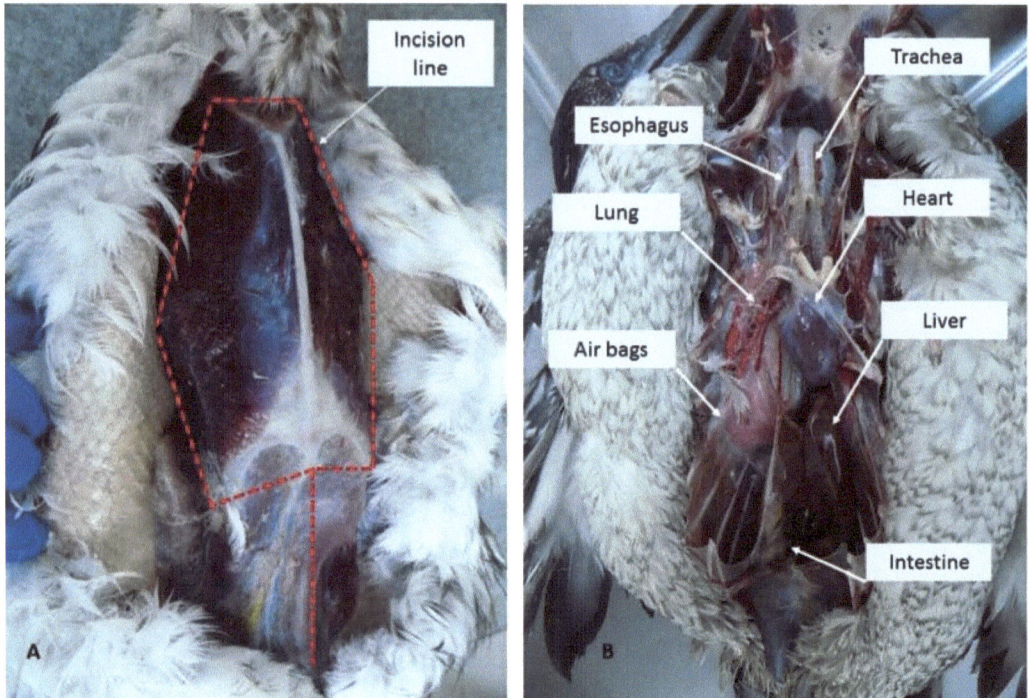

Fig. (19). A line of incision for the removal of the keel, in B internal organs *in situs* in the celomic cavity after incision.

Endocrine system and hematopoietic system

The thyroid and parathyroid are located cranially to the heart and lateral to the trachea, adjacent to the carotid arteries, bilaterally (Harcourt-Brown and Chitty, 2005; Chitty and Lierz, 2008; Chitty and Monks, 2018).

The normal thyroid glands are small, oval and with a reddish-brown colour. The parathyroid glands are very small and are often only identified with the aid of a magnifying glass or microscope. In juvenile and offspring the thymus, which lies laterally on both sides of the neck, should be observed cranial to the thyroid gland. This organ has a multilobulate appearance and is grey. Any change in the size, shape or colour of these organs should be recorded and the organs collected for further histological examination (Friend and Franson, 1999; Schmidt and Reavill, 2003; Dorrestein, 2008; Butcher and Miles, 2015).

In young animals, it is possible to observe the bursa of Fabricius. This lymphoid structure can be found on the dorsal wall of the cloaca and has a large lymph node-like appearance (Friend and Franson, 1999; Schmidt and Reavill, 2003; Dorrestein, 2008; Butcher and Miles, 2015). With the growth of animals, this organ becomes rudimentary, making it difficult to observe. Their identification in these animals can be facilitated by histological examination of cloacal tissue (Harcourt-Brown and Chitty, 2005; Chitty and Lierz, 2008; Chitty and Monks, 2018).

The spleen can usually be found in the dorsal area, at the angle between the ventricle and the proventriculus (Fig. **20**). Its volume is one-fourth to one-third the size of the heart, and its colouration in physiological situations is reddish-brown. The shape can vary with the species. It can have a round shape, namely in birds of prey and elongated in passerines (Friend and Franson, 1999; Schmidt and Reavill, 2003; Dorrestein, 2008; Butcher and Miles, 2015).

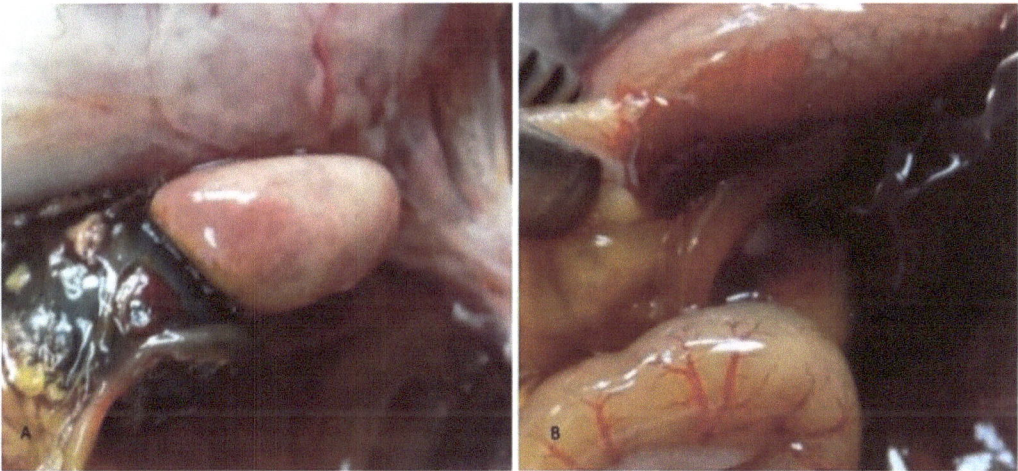

Fig. (20). A spleen in *Strix aluco* (bird of prey) and **B** is a *Caduerlis caduerlis* (passerine).

The adrenal glands are found cranial to the kidneys, and in the males are close to the gonads. They are two and have an orange colouration. They are removed together with the kidneys and gonads and should be observed *in situ* to assess their size, colour and presence of lesions (Friend and Franson, 1999; Schmidt and Reavill, 2003; Dorrestein, 2008; Butcher and Miles, 2015). During an infection, they may become hyperaemic or white. In animals that are subject to chronic stress they can be enlarged (Harcourt-Brown and Chitty, 2005; Chitty and Lierz, 2008; Chitty and Monks, 2018).

Cardiovascular System

In birds, the heart normally has a triangular shape and its size varies with the proportions of the animal. The pericardium should be transparent, thin, moisture and shiny and with an almost imperceptible amount of liquid. The organ should first be observed *in situ* to see if there is any change in size and shape. The heart is then removed from the coelomic cavity by cutting the large vessels (Friend and Franson, 1999; Schmidt and Reavill, 2003; Dorrestein, 2008; Butcher and Miles, 2015).

The pericardium must be observed to see if it presents changes of thickness, colouration or presence of liquid or blood in the pericardial sac. After removal of the pericardium, the surface of the myocardium should be observed. The section of the heart is made through an incision from the apex towards the atria so that it can observe the thickness of the myocardium and valves. Large vessels should be sectioned toward the bloodstream to visualize their lumen (Harcourt-Brown and Chitty, 2005; Chitty and Lierz, 2008; Chitty and Monks, 2018).

Digestive System and Annexed Glands

The evaluation of the digestive system is of particular importance, allowing the study of the diet of the animals (Harcourt-Brown and Chitty, 2005; Chitty and Lierz, 2008; Chitty and Monks, 2018). On the other hand, the detection of toxic substances or foreign bodies in food content can be fundamental in forensic cases.

Examination of the digestive system begins with the inspection of the oral cavity. It should be noted that the tongue and the beak have different shapes, sizes and mobility depending on the species and its diet. We must evaluate the presence of injuries such as wounds, plaques of fungus or parasites, foreign bodies, among others. In seabirds, the presence of wounds in the oral cavity may be indicative of the presence of accidentally ingested hooks (Friend and Franson, 1999; Schmidt and Reavill, 2003; Dorrestein, 2008; Butcher and Miles, 2015).

To remove the digestive system in a block, we should begin by making an incision in the muscles of the neck into the oropharynx to expose the trachea, oesophagus, and blood vessels. Afterwards, we remove the tongue, beginning by focusing on the commissure of the beak (Harcourt-Brown and Chitty, 2005; Chitty and Lierz, 2008; Chitty and Monks, 2018). Next, the hyoid bone and the pharyngeal tissues in the intermandibular region must be cut. With forceps, grasp the tongue and draw, so that the tongue, oesophagus, and large vessels are removed in a block. The trachea is sectioned distally under the syrinx (Friend and Franson, 1999; Schmidt and Reavill, 2003; Dorrestein, 2008; Butcher and Miles, 2015). The craw must then be elevated carefully to free it from the abdominal air sacs and to cut the mesenteric ligament. During this procedure, care must be taken to not damage the gonads and kidneys (Harcourt-Brown and Chitty, 2005; Chitty and Lierz, 2008; Chitty and Monks, 2018).

Some species belonging to the Picidae Order have an extremely long tongue, wrapped around the skull, which is exemplified in Fig. (**21**):

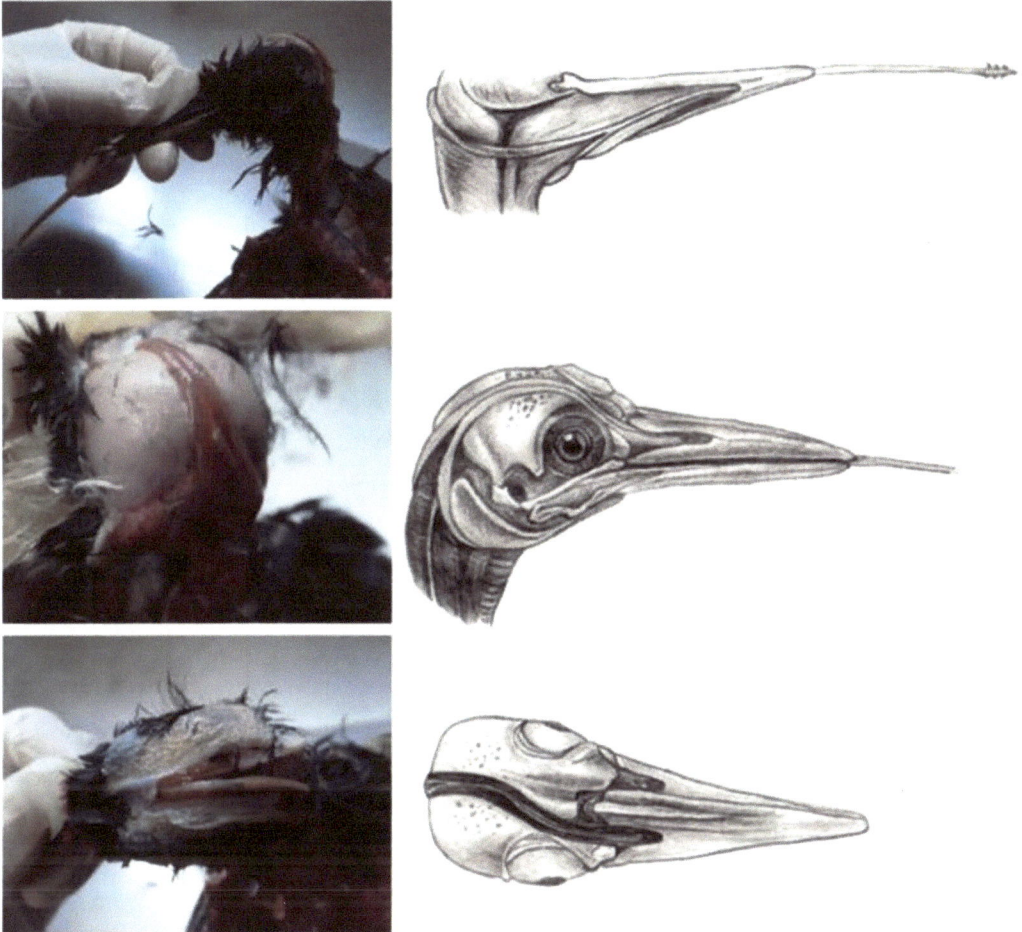

Fig. (21). Detail of the tongue on a *Picus viridis.*

The oesophagus, usually on the right side of the neck, has a thin wall. In birds have a greater distensibility when compared to other groups of animals (Harcourt-Brown and Chitty, 2005; Chitty and Lierz, 2008; Chitty and Monks, 2018).

In some bird species, the oesophagus exhibits a dilatation at the base of the neck, just at the entrance to the thorax, which is called craw. Its main function is food storage. In addition, in some species such as domestic geese, the craw is where it starts the chemical digestion of the grains due to the presence of amylase (Friend and Franson, 1999; Schmidt and Reavill, 2003; Dorrestein, 2008; Butcher and

Miles, 2015). When the craw has content, is possible to feel it by palpation. In Passeriformes and Anseriformes, it is less developed being the excess food stored in a pear-shaped dilation, whereas in insectivores and fructivores the craw has a fusiform form (Friend and Franson, 1999; Schmidt and Reavill, 2003; Dorrestein, 2008; Butcher and Miles, 2015).

Some orders as Sphenisciformes, Laridae and Caprimulgiformes do not have craw. For these birds, food is stored in the oesophagus or goes directly to the proventriculus. In the Columbiformes, this is a very well developed structure with two lateral chambers. It possesses glands that produce the "milk of the craw" that is used to feed the young. Other species, such as parrots, emperor penguins and finches also produce secretions in the craw (Harcourt-Brown and Chitty, 2005; Chitty and Lierz, 2008; Chitty and Monks, 2018). The *Columbia livia*, *Centrocercus urophasianus* and *Otis tarda* have an inflatable diverticulum that they use during the mating season as a resonance chamber. These structures must be sectioned in half after their extraction in order to allow the observation of their contents and lumen (Friend and Franson, 1999; Schmidt and Reavill, 2003; Dorrestein, 2008; Butcher and Miles, 2015).

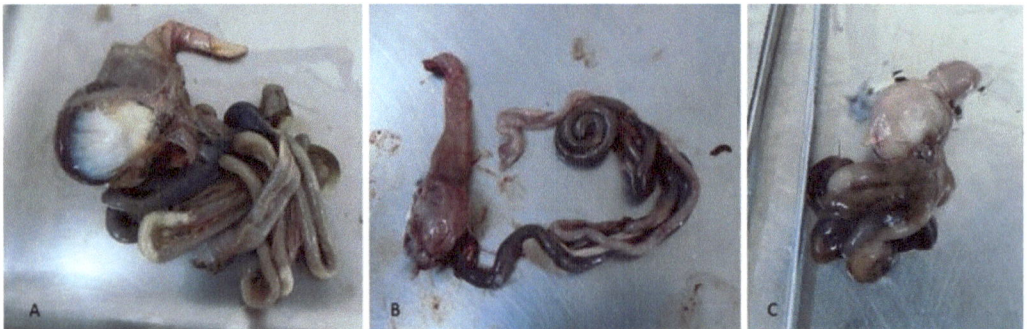

Fig. (22). Different types of digestive system, in granivorous **(A)**, carnivorous **(B)** and insectivorous **(C)** bird

Next, the stomach, consisting of the proventriculus and the ventricle (gizzard), divided by the isthmus (sphincter). In the first portion, the proventriculus, we can easily observe the gastric glands. Its morphology varies with species and diet. The ventricle (gizzard) is located to the left of the midline, caudal to the sternum, and possible to palpate (very firm consistency). It has a very developed muscular wall since it is responsible for the physical digestion of the fibrous food. Its inner mucosa is covered by a hard chitin membrane called the protective coiline layer, which has a bile-like colouration by the bile reflex of the duodenum. Often in this structure, we can find small stones that the birds consume to better digest the seeds mechanically (Harcourt-Brown and Chitty, 2005; Chitty and Lierz, 2008; Chitty and Monks, 2018).

In birds with a diet based on meat or fish, the proventriculus is the predominant portion, having the shape of a pear and a thin wall. In Sphenisciformes, for example, it can be extended caudally to the interior of the abdomen almost until reaching the cloaca (Friend and Franson, 1999; Schmidt and Reavill, 2003; Dorrestein, 2008; Butcher and Miles, 2015). The ventricle is poorly developed, serving only to hold the food long enough for the gastric juices to act. The grassy and omnivorous animals have a well-developed proventriculus. In species whose main diet is the fruit, like Lories, is little muscled, with a soft consistency (Fig. **22**) (Harcourt-Brown and Chitty, 2005; Chitty and Lierz, 2008; Chitty and Monks, 2018).

In the small intestine, the separation between the duodenum, jejunum and ileum is not very marked. The duodenum is U-shaped, with a proximal descending portion and ascending distal portion. In this U is located the pancreas. Under normal conditions, the pancreas possesses a pale colour and is composed by three lobes and three ducts connected to the duodenum (Friend and Franson, 1999; Schmidt and Reavill, 2003; Dorrestein, 2008; Butcher and Miles, 2015). This organ should be analysed, and samples collected as soon as possible, because it rapidly enters into autolysis, compromising its evaluation. The jejunum is separated from the ileum by the presence of Meckel's diverticulum, which is a remnant of the yolk sac. In nectarivorous species (*e.g.Trochilidae*) the small intestine is very short (Harcourt-Brown and Chitty, 2005; Chitty and Lierz, 2008; Chitty and Monks, 2018).

The intestine should be sectioned for evaluation of its contents and lumen. Often, parasites can be observed. In small birds, the intestine can be fixed in formaldehyde without opening, making small sections in the wall so that there is a better penetration by the formalin (Friend and Franson, 1999; Schmidt and Reavill, 2003; Dorrestein, 2008; Butcher and Miles, 2015).

The large intestine of the birds is quite short and has a larger diameter than the small intestine. This, together with the reproductive system and the urinary system, ends up in a chamber called a cloaca (Friend and Franson, 1999; Schmidt and Reavill, 2003; Dorrestein, 2008; Butcher and Miles, 2015). The cloaca is divided in three compartments (coprodaeum, urodaeum and proctodaeum) and has communication with the outside, where the waste of these systems is expelled (Harcourt-Brown and Chitty, 2005; Chitty and Lierz, 2008; Chitty and Monks, 2018).

Most species have two cecum, except the species of the Order Ardeidar. Others species only present a vestigial form like the Columbiformes and the Passeriformes. In animals of the Order Striginformes, the distal extremities are expanded and in the ostrich are long and sacculated. In some species, the cecum is not present, such as the case of some Coraciiformes, Piciformes and Psittaciformes (Harcourt-Brown and Chitty, 2005; Chitty and Lierz, 2008; Chitty and Monks, 2018).

It is also important to perform cytology of the proventriculus and intestine to observe the presence of parasites and bacteria. One example is megabatteries that affect many Passeriformes (Friend and Franson, 1999; Schmidt and Reavill, 2003; Dorrestein, 2008; Butcher and Miles, 2015).

The liver, in birds, is a brownish-coloured organ, bilobed, that is located around the heart (Harcourt-Brown and Chitty, 2005; Chitty and Lierz, 2008; Chitty and Monks, 2018). In neonates, it is normal to have a yellowish colour due to the mobilization of nutrients from the yolk sac. In some birds, such as parrots, the right lobe is larger giving it an asymmetrical appearance.

The liver should be removed by cutting the peritoneum and examined its size, consistency, colour, and other lesions present. In very fresh cadavers, blood from the cut of the hepatic vein can hide lesions in other organs. Therefore, the carcass must be first refrigerated before performing the *post-mortem* examination or remove the blood from the vein with a syringe (Friend and Franson, 1999; Schmidt and Reavill, 2003; Dorrestein, 2008; Butcher and Miles, 2015).

For a better examination of the liver parenchyma, several transverse incisions should be made in the organ and sectioned the present lesions to evaluating their depth.

If present, the gallbladder also should be examined for changes in its content, colour or thickness (Harcourt-Brown and Chitty, 2005; Chitty and Lierz, 2008; Chitty and Monks, 2018). Some species, like most Columbiformes, some Psitaccideos and Ostriches, do not have a gallbladder (Friend and Franson, 1999; Schmidt and Reavill, 2003; Dorrestein, 2008; Butcher and Miles, 2015).

Respiratory System

After the removal of the viscera referred to before, it is possible to see the lungs *in situ*. Air sacs are also part of this system but have already been observed previously.

As the oesophagus was previously removed, with some traction we were able to easily remove the trachea. The trachea should be sectioned with a pair of scissors, longitudinally, to observe its interior and mucosa (Friend and Franson, 1999; Schmidt and Reavill, 2003; Dorrestein, 2008; Butcher and Miles, 2015).

The trachea, in most birds, ends in a dilated structure called the syrinx. The syrinx is formed by the caudal rings of the trachea that merge into a cylindrical tympanum. This structure presents variations, as is the case of some species of the Order Anatidae in which the males have a cartilaginous structure surrounding the syrinx that is used as a resonance box during the mating season (Fig. **23**). In the case of swans, part of the trachea is inside the sternum, while in Galliformes and Cracidae the trachea emerges from the sternum and lies between the skin and pectoral muscles (Harcourt-Brown and Chitty, 2005; Chitty and Lierz, 2008; Chitty and Monks, 2018).

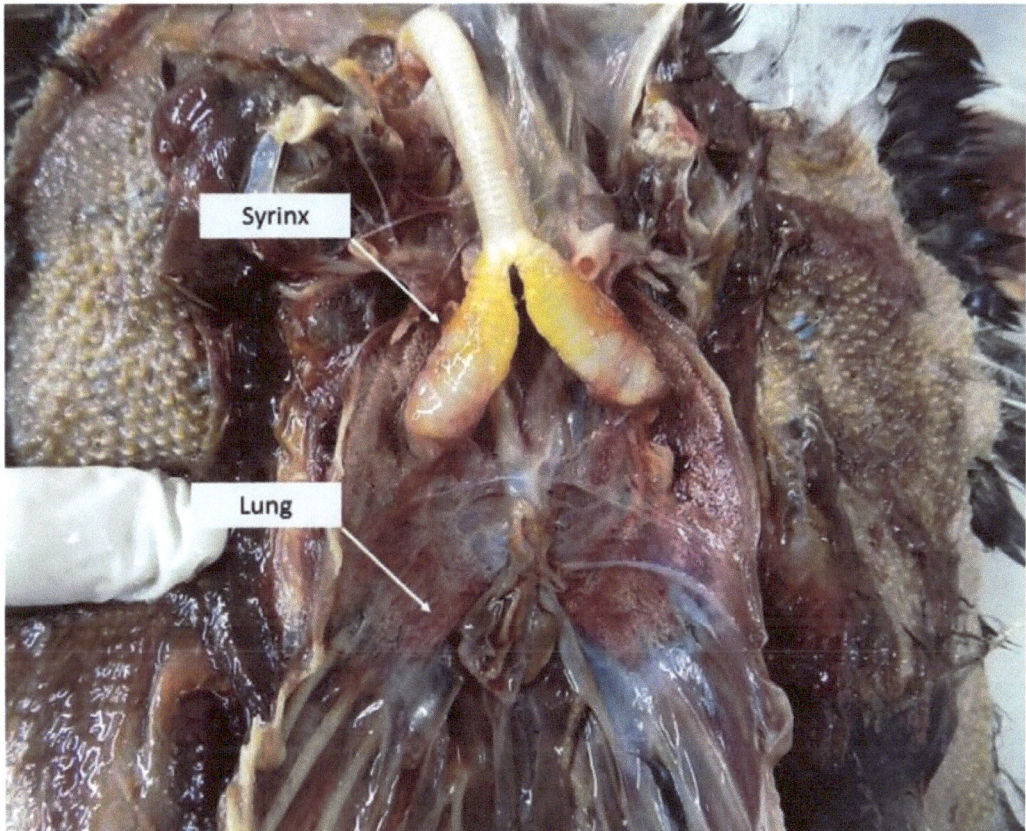

Fig. (23). Detail of the cartilaginous structures in the syrinx of a male *Melanitta nigra.*

The lungs are attached to the dorsal portion of the rib cage (Fig. **24**). The normal colour of the parenchyma of this organ is dark pink because it is quite irritated. They are stiffer than the lungs of mammals since they have a greater amount of cartilage (Harcourt-Brown and Chitty, 2005; Chitty and Lierz, 2008; Chitty and Monks, 2018). With forceps and a scalpel blade, they should be removed from the chest wall very carefully not to damage them. In this way, it is possible to observe lesions in the ventral portion of the lung and the thoracic wall. Then several incisions are made in the parenchyma to observe internally the parabronchi (Friend and Franson, 1999; Schmidt and Reavill, 2003; Dorrestein, 2008; Butcher and Miles, 2015).

Urinary System

The kidneys in the birds are in the renal fossa of the synsacrum (Fig. **24**). They have a reddish-brown colour, are bilateral and lobed. They are relatively large, extending from the caudal portion of the lungs to the caudal portion of the sacrificial sac. In most species, we can divide them into three parts: cranial, medial and caudal. In Passeriformes, it seems that the middle portion is absent, whereas in penguins and cranes have a caudal kidney fused in the midline (Friend and Franson, 1999; Schmidt and Reavill, 2003; Dorrestein, 2008; Butcher and Miles, 2015). The distinction between the cortex and the marrow is difficult to visualize in these animals. They do not have a urinary bladder. The ureters are also in pairs, ranging from the cranial lobe to urodeus, and, under normal conditions, are difficult to identify (Harcourt-Brown and Chitty, 2005; Chitty and Lierz, 2008; Chitty and Monks, 2018).

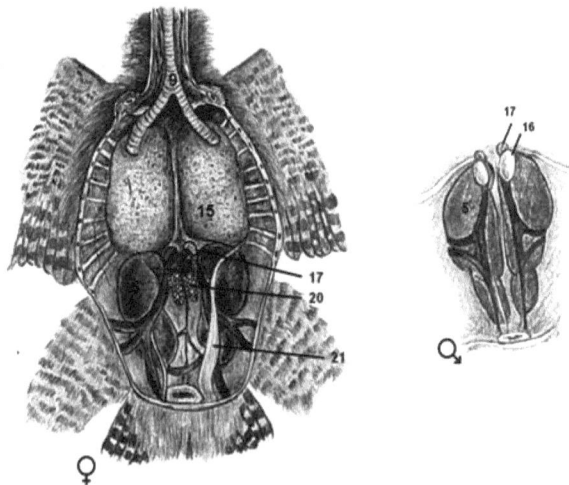

Fig. (24). Schematic representation of the renal and genital organs, male and female (5 - kidney, 9 - trachea, 10 - oesophagus, 15 - lung, 16 - testis, 17 - adrenal gland, 20 - ovary; 21- oviduct).

The kidneys are removed as a single unit. The outer and sciatic iliac blood vessels are laterally sectioned. Next, they can be detached from the dorsal wall of the sacrificial sac through blunt dissection, with the aid of a forceps and a scalpel, always being careful not to damage them (Friend and Franson, 1999; Schmidt and Reavill, 2003; Dorrestein, 2008; Butcher and Miles, 2015). Several cuts are made in the parenchyma to be able to observe possible lesions in depth. In very small birds they can be difficult to remove. As an option, you can remove all the synsacrum and fix in formalin (Friend and Franson, 1999; Schmidt and Reavill, 2003; Dorrestein, 2008; Butcher and Miles, 2015).

Genital System

The gonads are present in the cranial portion of the kidneys, near the adrenal glands (Fig. **24**). Their observation allows determining the sex of the animal, although in some situations this is not possible, *e.g.* advanced autolysis or newborn animals. They should be observed *in situ* before they are removed (Friend and Franson, 1999; Schmidt and Reavill, 2003; Dorrestein, 2008; Butcher and Miles, 2015).

In females, usually, the left ovary and oviduct are the most developed, except in the species of Falconiformes of the family Cathartidae, Accipitridae and Falconidae and *Apteryx australis*, which have well developed both ovaries (Harcourt-Brown and Chitty, 2005; Chitty and Lierz, 2008; Chitty and Monks, 2018).

Depending on the time of year and age of the animal, the gonads undergo changes. In juvenile birds, the ovary is small, with a triangular shape, greyish with an irregular surface and a fine oviduct (Harcourt-Brown and Chitty, 2005; Chitty and Lierz, 2008; Chitty and Monks, 2018). The follicles in adult females during the reproduction season have a yellowish colouration, but in some species, they can be pigmented, with different sizes and the oviduct is thickened. The ovary should be observed for the presence of follicles and eventual changes. The oviduct should be sectioned with scissors to access to the lumen (Fig. **25**) (Friend and Franson, 1999; Schmidt and Reavill, 2003; Dorrestein, 2008; Butcher and Miles, 2015).

Male gonads, in birds, are found internally, in pairs, cranial to the kidney, near the adrenal glands. They usually have an elongated to cylindrical ovoid shape and maybe pigmented black in some species or have a light yellow colour. In juvenile birds they have a yellow colouration due to the presence of interstitial lipid cells and the left testis is more developed (Fig. **25**) (Friend and Franson, 1999; Schmidt and Reavill, 2003; Dorrestein, 2008; Butcher and Miles, 2015). As in females, the gonads vary in size according to the breeding season. In the breeding season, they grow in size. The penis is absent or vestigial in most species except Ratites and Anseriformes (Harcourt-Brown and Chitty, 2005; Chitty and Lierz, 2008; Chitty and Monks, 2018).

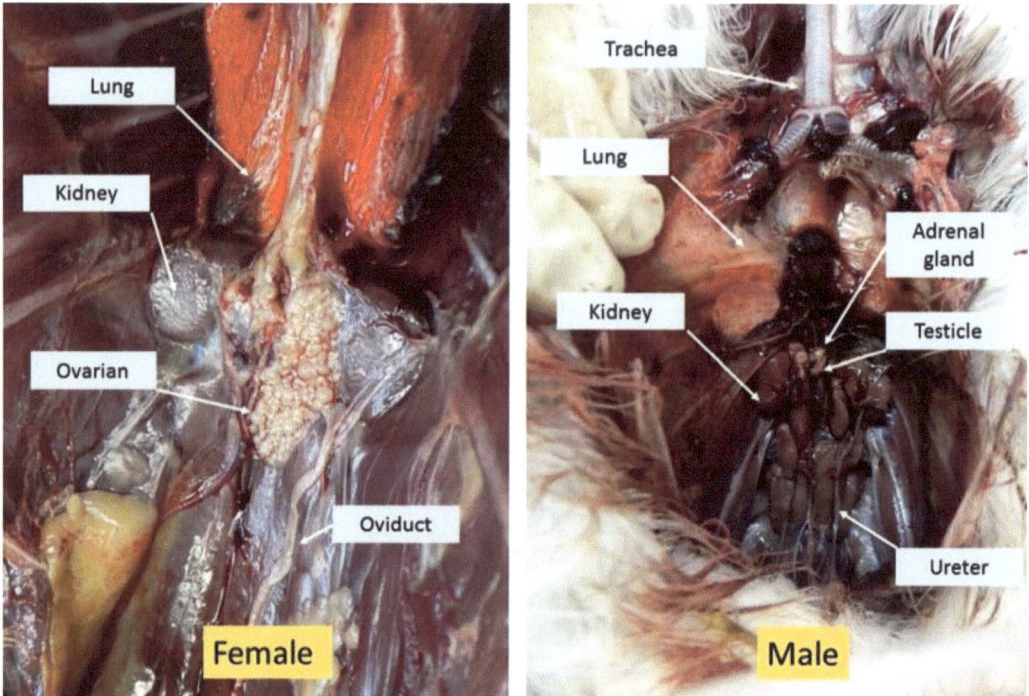

Fig. (25). Schematic representation of the renal and genital organs, male and female.

Nervous System

The evaluation of the nervous system is often neglected during the necropsy of a bird. However, it can provide valuable information for the diagnosis of certain diseases or to diagnose acute death from the traumatic origin (Harcourt-Brown and Chitty, 2005; Chitty and Lierz, 2008; Chitty and Monks, 2018).

The brachial plexus can be seen in the brachial region before the viscera are removed. The lumbosacral plexus should be observed when there is suspicion of paralysis in the posterior limbs and it is located below the kidneys, being possible to be observed after their removal (Friend and Franson, 1999; Schmidt and Reavill, 2003; Dorrestein, 2008; Butcher and Miles, 2015).

The fixation of the nerves in formalin should be done with a piece of cardboard to maintain their position and avoid possible losses.

Access to the brain is relatively easy. Begin by removing the feathers from the head, then make a sagittal incision in the scalp and remove the skin. Before opening the cranial cavity, haemorrhages or hematomas in the cranial bone should be observed (Friend and Franson, 1999; Schmidt and Reavill, 2003; Dorrestein, 2008; Butcher and Miles, 2015). They may be due to trauma, agonizing death or cadaveric phenomena. To access the brain, a sagittal incision should be performed on the calvarium, using scissors, and with the help of scissor or forceps to remove it. The skull is placed downwards and gently removes itself with the aid of gravity (Figs. **26-28**) (Harcourt-Brown and Chitty, 2005; Chitty and Lierz, 2008; Chitty and Monks, 2018).

If spinal cord injury is suspected, the portion thought to be altered may be removed and placed in formaldehyde as a whole or sectioned to the middle with a pair of scissors or bone saw to observe the bone marrow and medullary canal (Friend and Franson, 1999; Schmidt and Reavill, 2003; Dorrestein, 2008; Butcher and Miles, 2015).

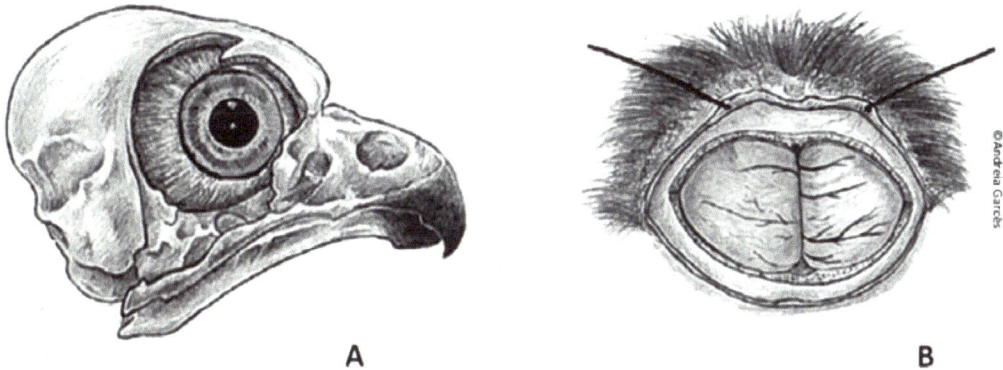

A   **B**

Fig. (26). Schematic representation of the skull after removal of skin **(A)** and exposed brain after removal of part of the skull **(B)**.

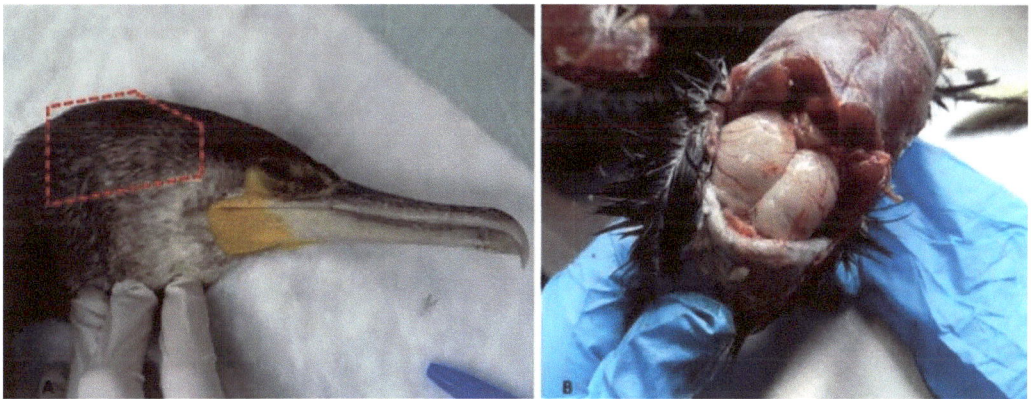

Fig. (27). Schematic representation of the skull after removal of skin **(A)** and exposed brain after removal of part of the skull **(B)** (*Phalacrocorax carbo*).

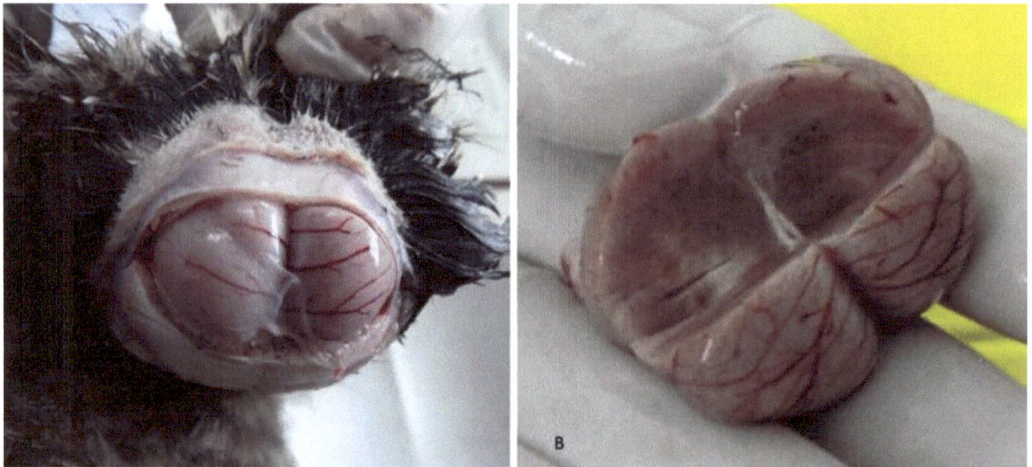

Fig. (28). Brain exposed after removal of part of the cranial box in a *Strix aluco***(A)**, and the exposed brain in **(B)**.

Locomotive System

The skeleton has already been partially evaluated by palpation of limbs. During the opening of the carcass, we should also observe some injuries such as haemorrhages in keel bones, bruises, rib fractures or other bones (Harcourt-Brown and Chitty, 2005; Chitty and Lierz, 2008; Chitty and Monks, 2018) such as coracoid, hyoid, thyroid, which are difficult to feel through palpation and even through complementary exams such as radiography. We must also section the joints to observe if they present any changes such as inflammation, fractures or foreign bodies (Friend and Franson, 1999; Schmidt and Reavill, 2003; Dorrestein, 2008; Butcher and Miles, 2015).

REFERENCES

Brooks, J W (2018) *Veterinary Forensic Pathology.* Springer International Publishing, NY, USA. [http://dx.doi.org/10.1007/978-3-319-67172-7]

Butcher, GD & Miles, RD (2015) *Avian Necropsy Techniques.* Gainesville, Florida.

Chitty, J, Lierz, M (2008). *Manual of raptores, pigeons and waterfowl.* British Small Animal Veterinary Association, UK.

Chitty, J, Monks, D (2018). *BSAVA Manual of Avian Practice: a foundation manual.* British Small Animal Veterinary Association, UK.

Dorrestein, GM (2008) Clinical pathology and post-mortem examination. *Manual of raptors, Pingeons and Passerine Birds* British Smalll Animal Veterinary Association, London 73-93. [http://dx.doi.org/10.22233/9781910443101.9]

Friend, M & Franson, JC (1999) *Field Manual of Wildlife Diseases - General Field Procedures and Diseases of Birds.* Library of Congress, Cataloging, USA.

Garcês, A & Pires, I (2017) *Manual de Técnicas de Necrópsia em Animais Selvagens.* Arteology, Porto.

Harcourt-Brown, N, Chitty, J (2005). *BSAVA Manual of Psittacine Birds.* British Small Animal Veterinary Association, UK.

King, JM (2014) *The necropsy book: A Guide for Veterinary Students, by The Necropsy Book.* Charles Louis David DVM Foundation Publisher, Ithaca.

King, JM, Roth-Johnson, L, Dodd, DC & Newsom, ME (2013) *The necropsy book: A Guide for Veterinary Students, Residents, Clinicians, Pathologists, and Biological Researchers.* College of Veterinary Medicine, Cornell University, NY, USA.

Munson, L (2000) *Necropsy of Wild Animals.* Wildlife Health Center, School of Veterinary Medicine, USA.

Schmidt, R E & Reavill, D R (2003) *A Practitioner's Guide to Avian Necropsy.* Zoological Education Network, Lake Worth, Florida.

CHAPTER 5

Necropsy in Wild Mammals

Outline: In this chapter, we describe the method of necropsy in wild mammals, offering some information regarding the different orders and anatomic characteristics of the determined species.

Keywords: Animals sentinels, Carnivore, Conservation, Marine Mammal, Mortality, Necropsy, Pathology, *Post-Mortem*, Terrestrial, Wild Mammals.

GENERAL CONSIDERATIONS

A *post-mortem* examination in mammals should be performed as soon as possible after death. If it is not possible until 72 to 96 hours after death, the animal should be refrigerated. To improve the process, in small to medium-sized carcasses, the fur should be wet with cold water and placed in a plastic bag in the refrigerator (4°C). In larger cadavers, refrigeration does not occur fast enough to prevent autolysis of organs such as the intestine, pancreas, or kidneys. In this case, the necropsy must be performed immediately. If this is not possible, an incision should be made in the abdomen before cooling the carcass. If the procedure cannot be completed until 96 hours after death, the corpse should be frozen immediately, despite the numerous damages resulting from this preservation process (Fig. **1**) (Munson, 2000; Woodford, Keet and Begins, 2000).

In addition to the materials and equipment described above, some species are extensive, such as horses, ursids or elephants, it is necessary to have a large necropsies room with structures such as cranes or ropes to raise the animal.

The identification of the species should be made prior to the necropsy. Knowledge of the status of the species, for example, if it is protected or hunting, is important for a possible warning from the competent authorities.

Andreia Garcês & Isabel Pires

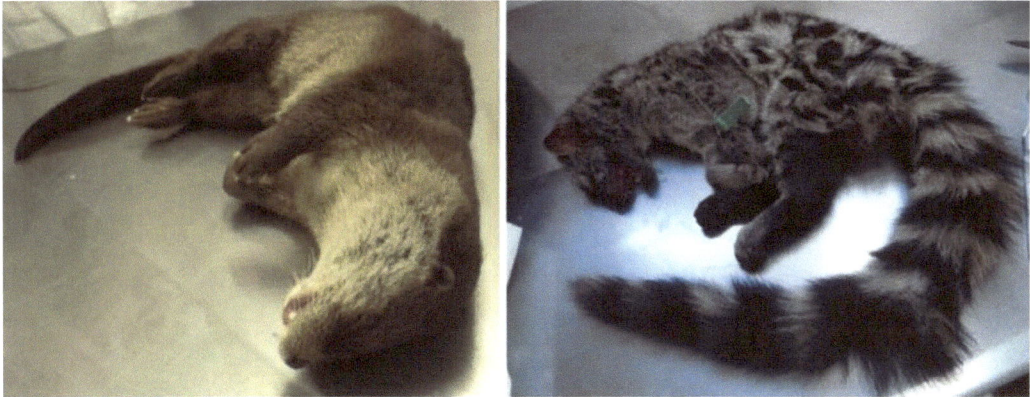

Fig. (1). *Lutra lutra* and *Genettta genetta,* both run over by a motorized vehicle.

The knowledge of the behaviour, diet and territory of each animal can aid in the interpretation of the observations in the necropsy and contribute to the diagnosis of the cause of death. Before performing the necropsy, an x-ray may be useful to rule out suspected firearms, foreign bodies or fractures (Somvanshi and Rao, 2009; King *et al.*, 2013; McDonough and Southard, 2017).

EXTERNAL EXAMINATION

The necropsy begins before the carcass incision. Careful observation of the animal corpse can provide vital information about the cause and mechanism of animal death or previous diseases. The degree of preservation of the carcass is one of the first observations. Thus, cadaveric phenomena must be observed to estimate the time of death. Before alterations are detected, the animal's typical characteristics, such as body condition, gender and age, should be noted (King *et al.*, 2014).

Table 1. Body condition in wild mammals.

Mammals Body Condition			
	Fat	**Bones**	**Health**
CACHECTIC	None	Very prominent	No
SKINNY	Very little	Prominent	Yes
NORMAL	Some	Palpable	Yes
FAT	Very	No palpable	Yes

The animal should be weighed, and its body condition assessed. The assessment of body condition is a parameter that is somewhat subjective and easily maskable by the coat. To facilitate, we can divide into four categories: cachectic, thin, normal and fat. The characteristics of these categories are described in Table **1** & (Fig. **2**). One should not forget to associate with the season of the year because some animals hibernate, like bears, European hedgehogs or bats. Therefore, it will be standard in autumn to find fat animals with vast fat reserves, while in early spring, when they leave hibernation, lean with little or no adiposity reserves (Kenneth and Rakich, 1994; Somvanshi and Rao, 2009; McDonough and Southard, 2017).

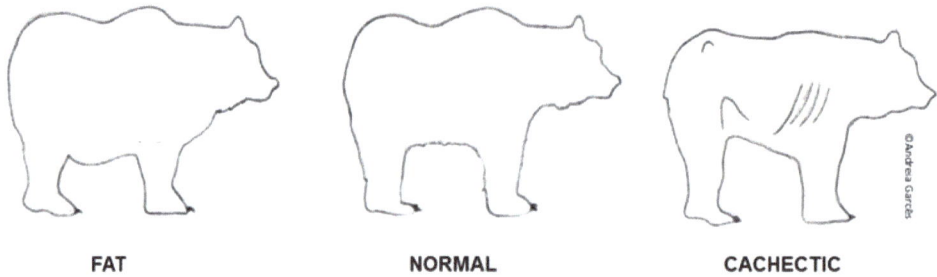

Fig. (2). Scheme of body condition in mammals.

It is essential before starting the procedure to remove all possible biometric measures. These are not only important for the study of the species but can give us information about the sex, age or subspecies of the animal (Fig. **3** & **4**) (King *et al.*, 2014).

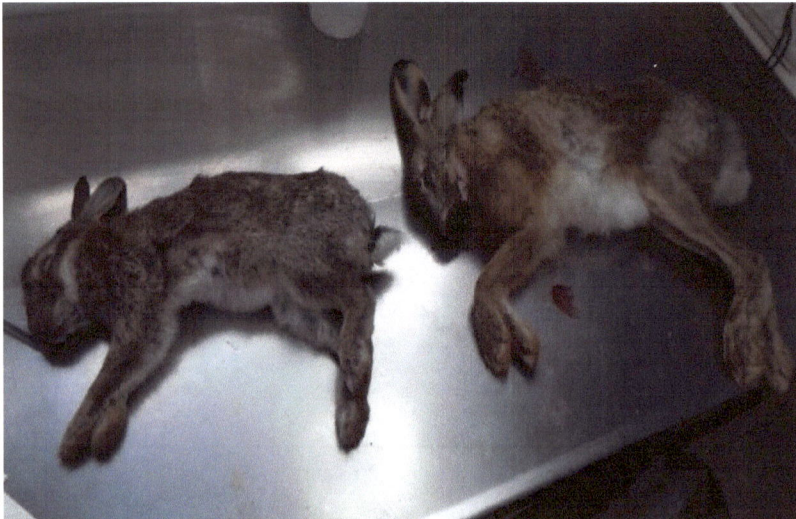

Fig. (3). *Oryctolagus cuniculus* and *Lepus europaeus*.

Fig. (4). Representation of mammal species from different orders: **A**-*Equus ferus przewalskii*; **B**-*Pipistellus pipistrellus*; **C**-*Genetta genetta*; **D**-*Sus scrofa* ; **E**-*Mustela putoris*; **F**-*Cervus elaphus*; **G**-*Vulpes vulpes*; **H**-*Loxodante africana*; **I**-*Physeter macrocephalus*; **J**-*Cynomys gunnisoni*.

The sexing of animals, in mammals, is, in general, natural, since they have external genitalia that allows to identify them. However, in some species, such as the hedgehog (*Erinaceus europaeus*), the testes have an intra-abdominal location. However, it is possible to visualise, in adult males, a prominent prepuce in the midline, about 5 cm cranial to the anus, whereas in adult females the vagina is positioned about 1 cm from the anus (Fig. **5**) (Cardoso, 2002; McDonough and Southard, 2017).

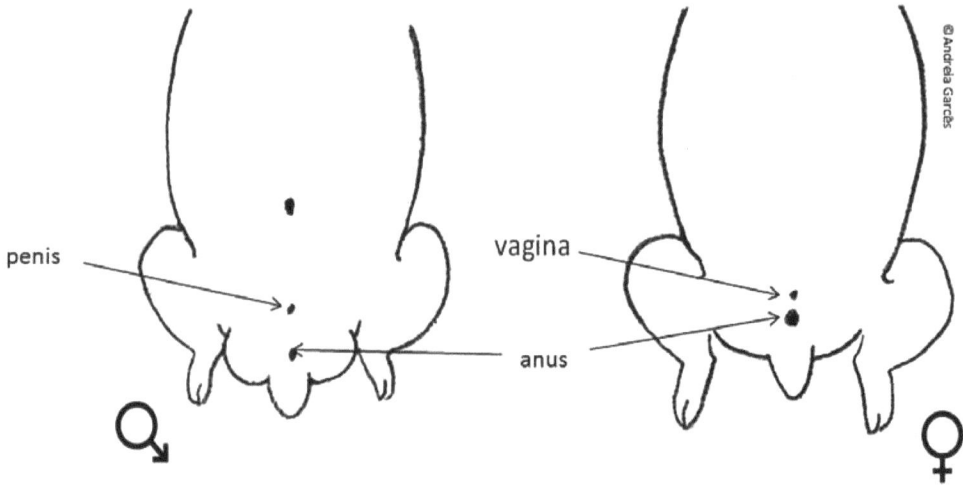

Fig. (5). Representation of sexing in rodents (male on the left, female on the right).

Concerning age, the animals should also be categorised, if possible, according to their age: calves, young, adult or geriatrics.

The presence of identifying marks such as tattoos, microchips or the presence of apparatuses such as radio transmitter tags should be detected.

The evaluation of the external habit includes visual examination and palpation. Through visual inspection, changes in body shape should be assessed (Fig. **6**). Other external features such as scars, characteristic colourations on the coat, changes in the coating, such as albinism or melanism, are pointed out. In species whose coat has patterns of stripes or pints make a sketch or photos of the pattern because these marks are unique in each animal and allow its identification (Friend and Franson, 1999).

The condition of the coat should also be recorded: if it is wet, dirty stools or substances such as blood, oils, oil, burnt, among others. In the case of foreign substances on the hair, some samples of hair should be collected for further analysis for identification. If ectoparasites are present, they should be collected and later identified. The observation of the nails/claws is also essential, as their sharp growth or wear and tear give us hints as to whether the animals were in captivity (Kenneth and Rakich, 1994; Cardoso, 2002; King *et al.*, 2014).

Fig. (6). Representation of the head of mammals of different orders: **A**-*Equus ferus przewalskii*; **B** –*Tadarida teniotis*;**C** –*Meles meles;***D** –*Bison bonasus*; **E**-*Panthera pardus*; **F**-*Capreolus capreolus*, that show the different adaptations to the diverse diets.

In the palpation of the carcass, dislocations, fractures, amputations, wounds, foreign bodies, emphysema, subcutaneous masses, muscular atrophies, dermatitis, zones of alopecia, dehydration or deformations can be observed.

All-natural orifices should be systematically inspected: eyes, ears, nostrils, oral cavity, anus, and genitals (Figs. **7-12**). The colouration, brightness, presence of erosions, ulcers, exudates, blood, foreign bodies, parasites, or masses should be recorded. The observation of the mucous membranes may be more difficult in some species. For instance, to observe the oral cavity in animals such as *Sus scrofa* or rodents it may be necessary to perform an incision on the labial commissure and cut the masseter muscle (Fig. **7**) (Cardoso, 2002; King *et al.*, 2014).

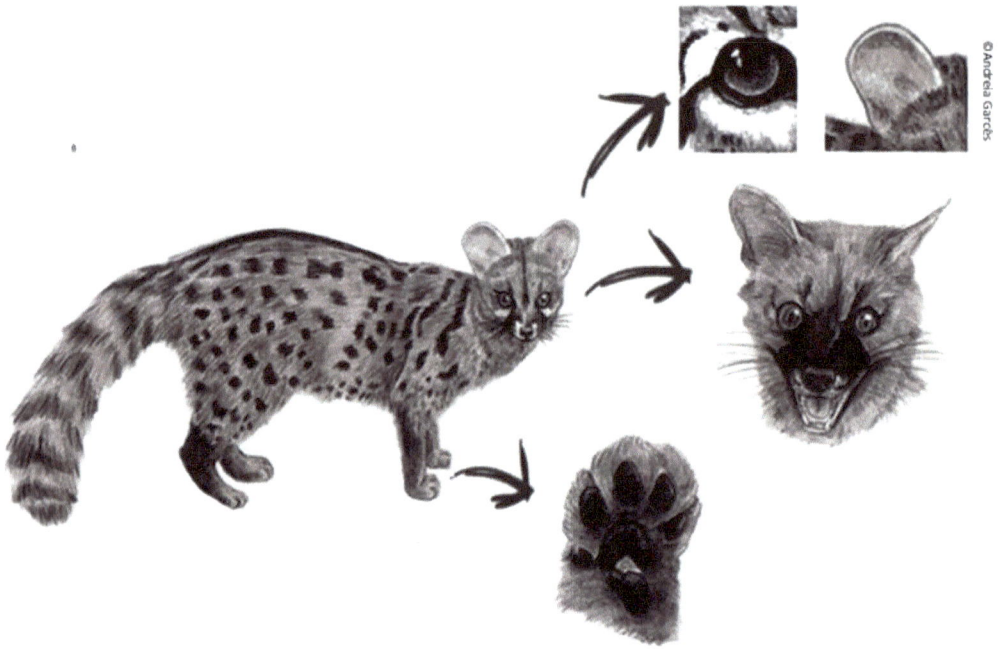

Fig. (7). External examination in detail of the eye, ear, limb and oral cavity of a carnivore (*Genetta genetta*).

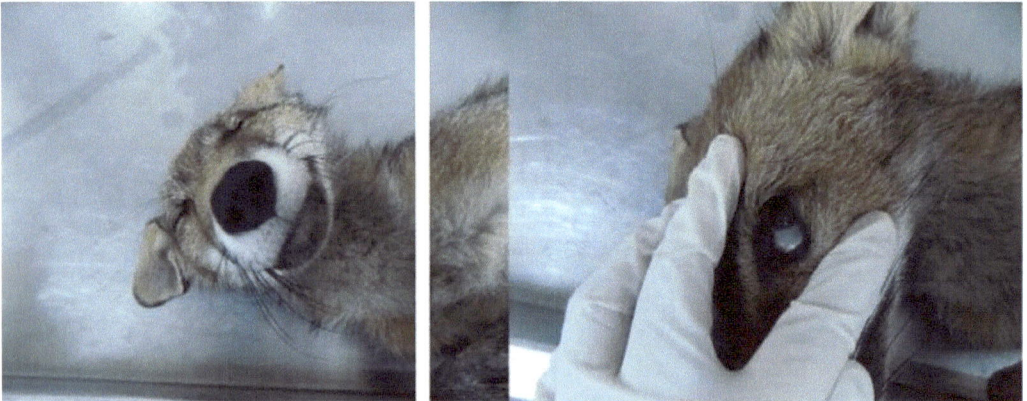

Fig. 8 cont.....

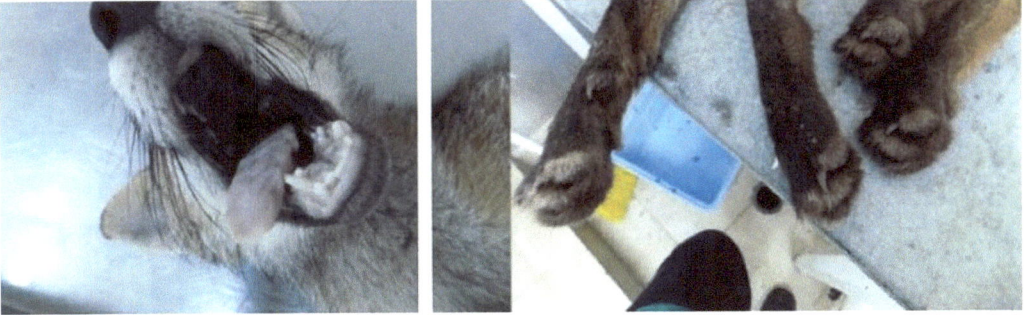

Fig. (8). External exam in a *Vulpes vulpes.*

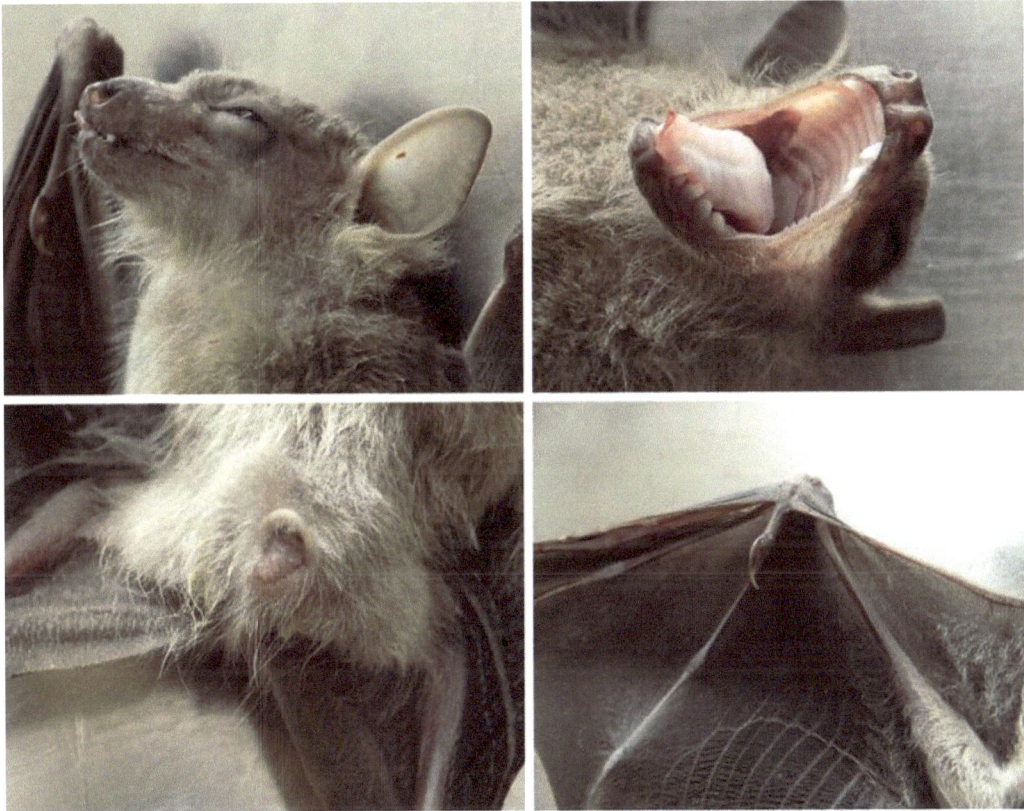

Fig. (9). External exam in a *Rousettus egyptiacus.*

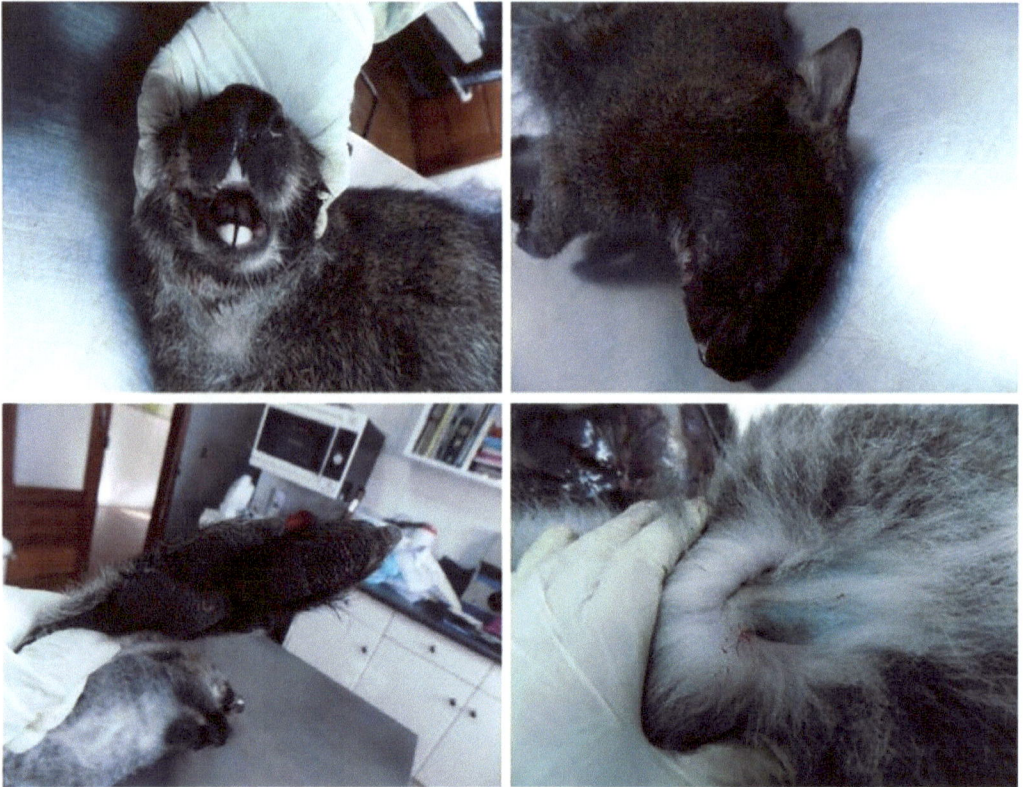

Fig. (10). External exam in a *Macropus rufagriseus*.

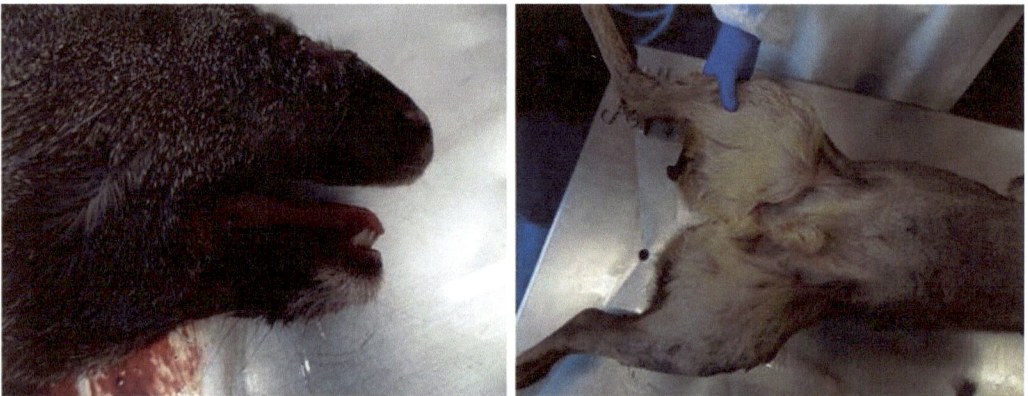

Fig. 11 cont.....

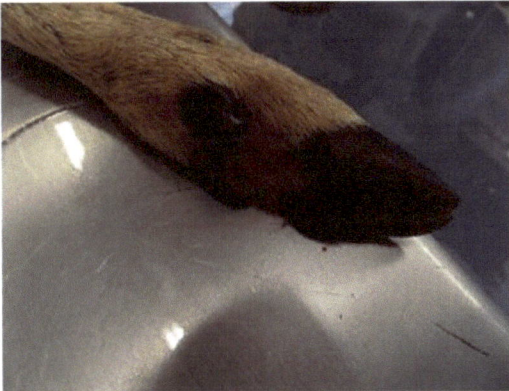

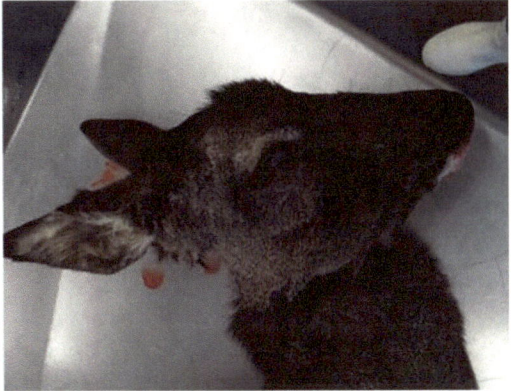

Fig. (11). External exam in a *Capreolus capreolus*.

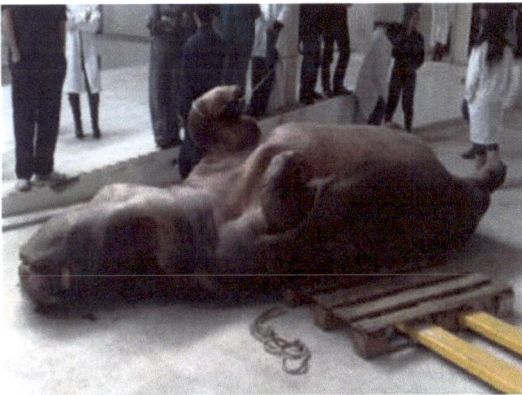

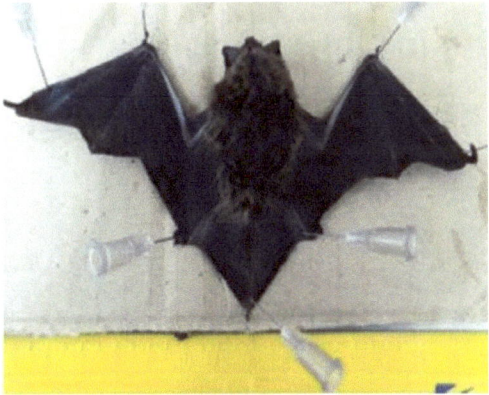

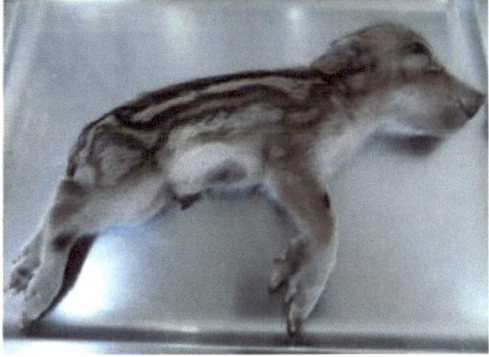

Fig. 12 cont.....

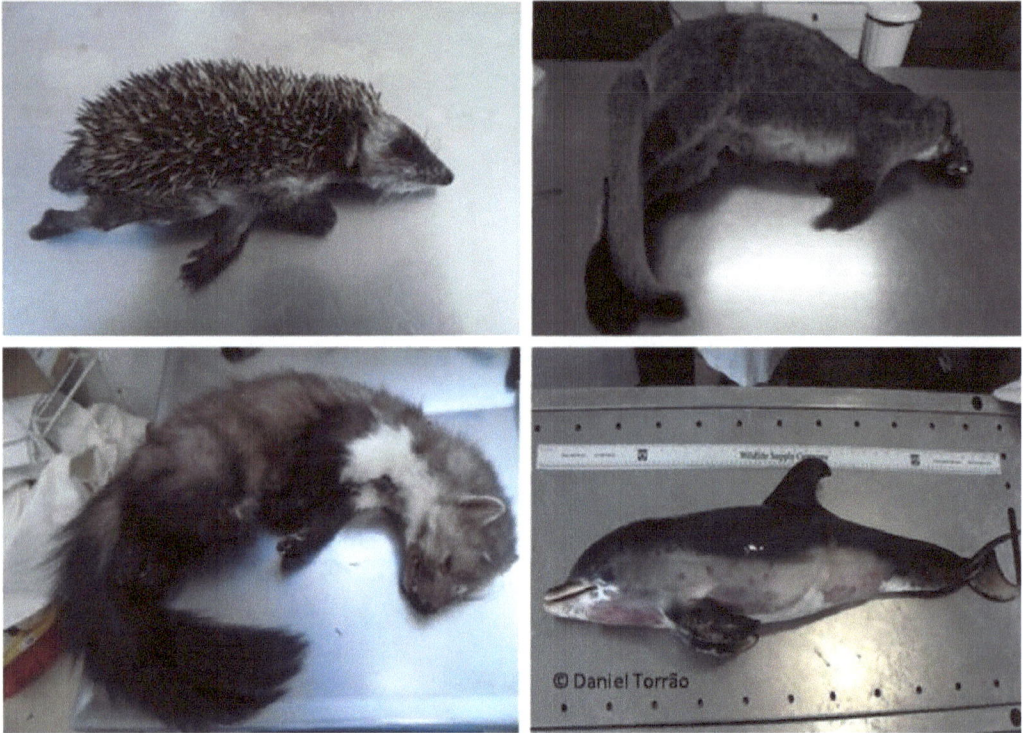

Fig. (12). Examples of an external examination of mammals of different orders.

The colouration of mucous membranes may indicate pathological processes such as inflammation if congested or severe anaemia if pale; cardiac problems when cyanotic or jaundice when the colour is yellow.

In addition to the oral mucosa, there are changes in the tongue such as trauma or ulcers, and later, multiple cross-sections are performed.

In the teeth, the dental formula should be pointed out and compared with what is described for the species in question. The teeth are observed to detect possible losses, periodontal disease, accumulation of tartar, gingivitis, wear, overgrowth (prevalent in rodents in captives with incorrect diets), among other changes. In cubs, possible congenital changes such as cleft palate or cleft lip could be detected (Munson, 2000; Cardoso, 2002; Wobeser, 2006).

If the degree of decomposition is not very advanced, the eyes should be observed. We can evaluate the presence of haemorrhages, masses, ulcers, cataracts, foreign bodies, exophthalmos, keratoconjunctivitis, among other lesions. If the carcass is very fresh, it is even possible to look into the back of the eye with an ophthalmoscope.

Palpation of the submandibular, pre-scapular and popliteal lymph nodes is essential. Its increased size may be indicative of systemic or localised inflammation (Munson, 2000; King *et al.*, 2013).

INTERNAL EXAM

The hair may be soaked in water or disinfectants as 70% alcohol, to reduce the release of debris that may have zoonotic agents.

The position of the carcass for the procedure depends on its size. The large ones should be positioned laterally, except for primates, since it is easier to access the viscera in the dorsoventral position. In this position, we can see physiological variations such as gestation. In single-digit ungulates (Equidae and Perissodactyla), this position allows better manipulation of the cecum and colon. In ungulates of two digits (Artiodactyla), this position allows the rumen to not block access to the remaining abdominal viscera (Fig. **13**) (King *et al.*, 2013).

Next, an incision should be made, deep in the armpit area and in the groin of the limbs at the top (Fig. **14**). The incision should be deepened until both limbs are disarticulated and placed back to allow access to the abdomen and chest (Fig. **14**) (King *et al.*, 2013).

In the case of marine animals due to the thickness of their skin, a midline incision and several sagittal cuts with a spacing of about 25cm should be made, to facilitate the removal of the skin (Fig. **15**) (Mclellan, Rommel and Pabst, no date; Raverty, Columbia and Gaydos; Cramer, Ketten and Montie, 2007; Eros *et al.*, 2007)

Carcasses of small mammals such as bats or small rodents can be attached to a board to facilitate the procedure. A small to a medium-sized corpse, such as *Erinaceus europaeus* or *Vulpes vulpes*, may be placed in dorsoventral position. To stabilise these animals for the post-mortem examination, it is necessary to focus on the internal face of the four limbs until the joints are involved. These should be disarticulated so that the limbs fall to the side only connected by soft tissues, and in this way, stabilise the carcass. The coxo-femoral joint should be observed, and any changes must be recorded.

Fig. (13). Exemplification of incision methods in a *Cervus elaphus*.

Fig. 14 cont.....

Fig. (14). Exemplification of incision methods in a *Vulpes vulpes*.

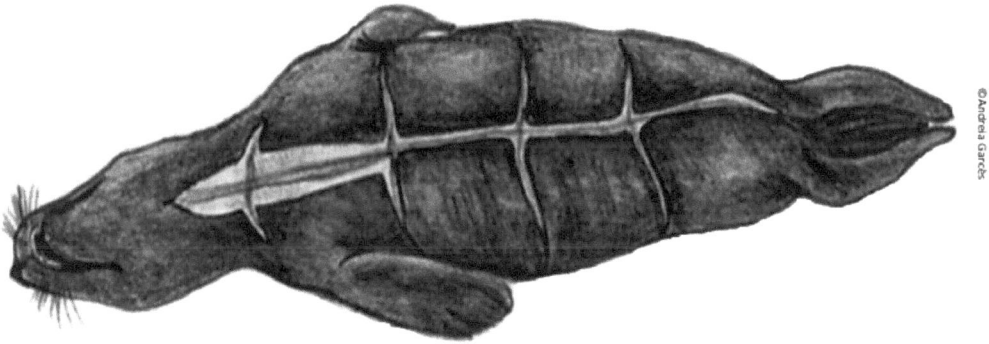

Fig. (15). Exemplification of incision methods on a *Hydrurga leptonyx*.

One of the problems, at the beginning of the necropsy, because the fur wears the edge of the knives quickly. Therefore, the edge of the blade should only have contact with the fur when making the incisions in the limbs. Then the skin should be moved away to place the knife in the subcutaneous tissue with the back of the blade facing the animal. Cut yourself, from the inside out, reducing the wear of the knife (King *et al.*, 2013).

The incisions made in the limbs should be extended toward the midline. From there, from the mandibular symphysis, it flows caudally to the perineum, close to the genitalia. After the midline incision is complete, the skin is removed, and the subcutaneous tissue is exposed. Therefore, we can observe the presence of haemorrhages, bruises or wounds, which are often masked by the thick coat. At this point, perforation of the abdominal cavity should be avoided. In this stage is possible to observe the cervical, pre-scapular, axillary, inguinal and popliteal lymphatic nodules (Fig. **16**) (Munson, 2000; King *et al.*, 2013; McDonough and Southard, 2017).

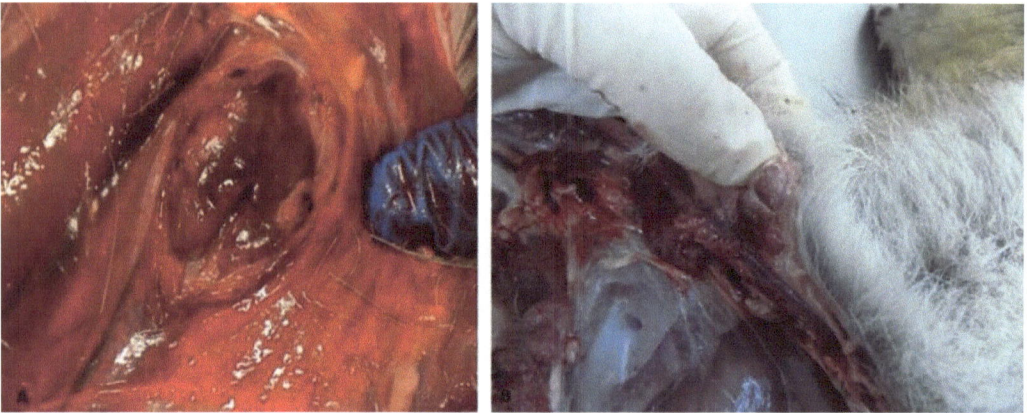

Fig. (16). A is presented an incision in the pre-scapular lymphatic nodule of a *Capreolus capreolus* and **B** in a *Vulpes vulpes.*

In females, we should observe the mammary gland. This can indicate the physiological state as gestation, pseudo gestation, or lactation. It is important to palpate this gland to detect the presence of mastitis, masses, or abscesses. Subsequently, the mammary gland should be removed.

To access the abdominal cavity, an incision should be made along the midline caudally toward the pubic bone and cranially towards the diaphragm/sternum. During this step, perforation of the gastrointestinal organs should be avoided, which would contaminate the rest of the organs and difficult the observation of any lesions (Figs. **17-20**).

The incision of the transverse abdominal muscles below the sternum and parallel to the last rib continues, continuing at the level of the spine. The diaphragm is then examined and drilled to assess the negative pressure on the thorax. With a pair of scissors or saw, depending on the size of the animal, the costochondral junctions are cut, the tip of the xiphoid process, then elevated the sternum, exposing the thoracic cavity (King *et al.*, 2013).

The anatomical position of the organs is observed in the thoracic and abdominal cavities. It is evaluated for the presence of free liquid, adhesions, parasites, cysts, tumours, inflammations, or other lesions. The amount of abdominal and mesenteric fat deposits present is also evaluated,

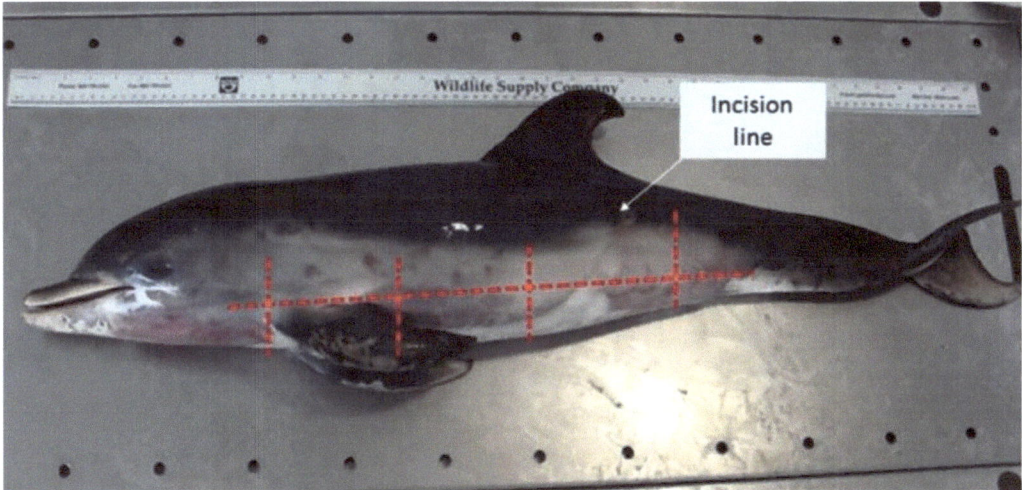

Fig. (17). Exemplification of the incision line in a *Tursiops truncatus*, (photo provided by Daniel Torrão).

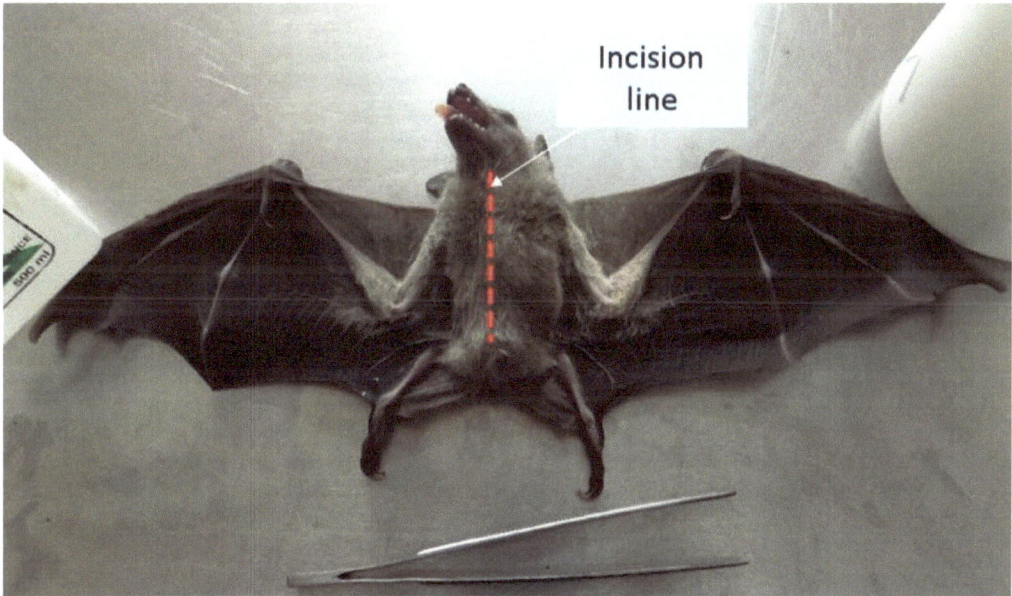

Fig. 18 cont.....

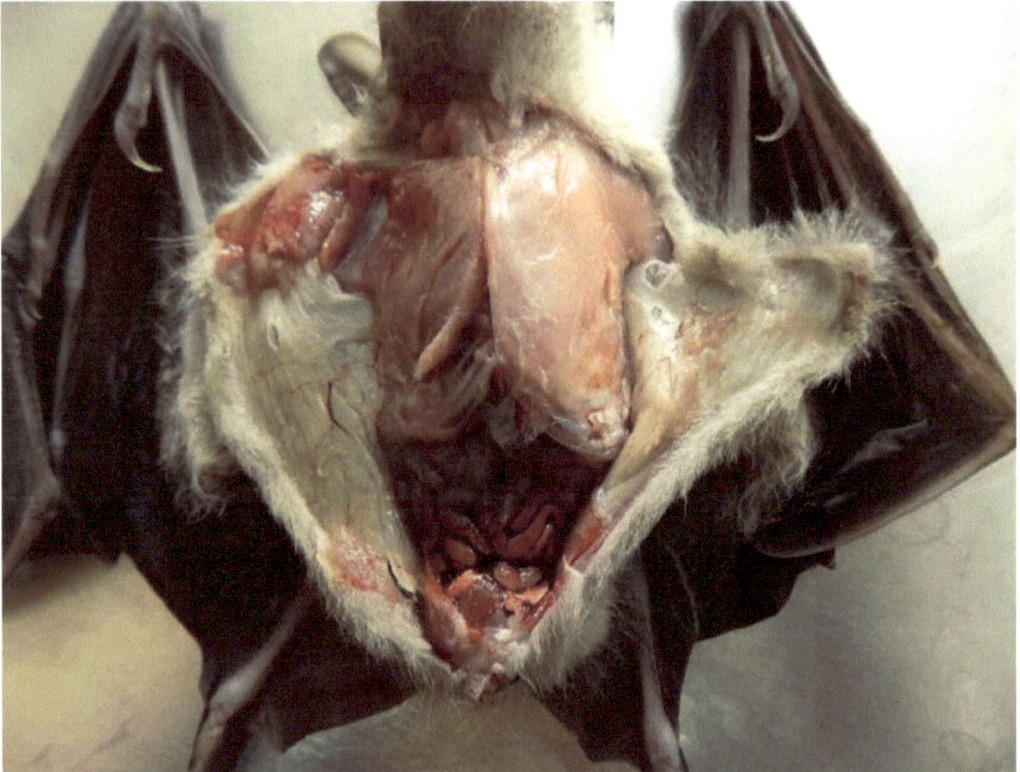

Fig. (18). Exemplification of the incision line and exposition of the pectoral muscles in a *Rousettus egyptiacus*.

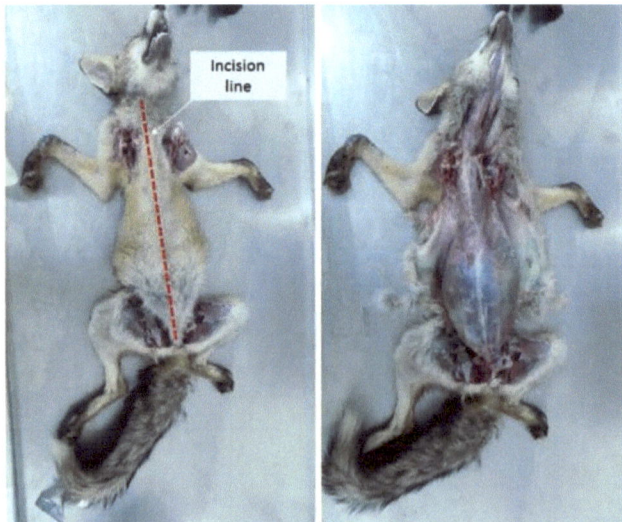

Fig. (19). Exemplification of the incision line and exposition of the pectoral muscles in a *Vulpes Vulpes*.

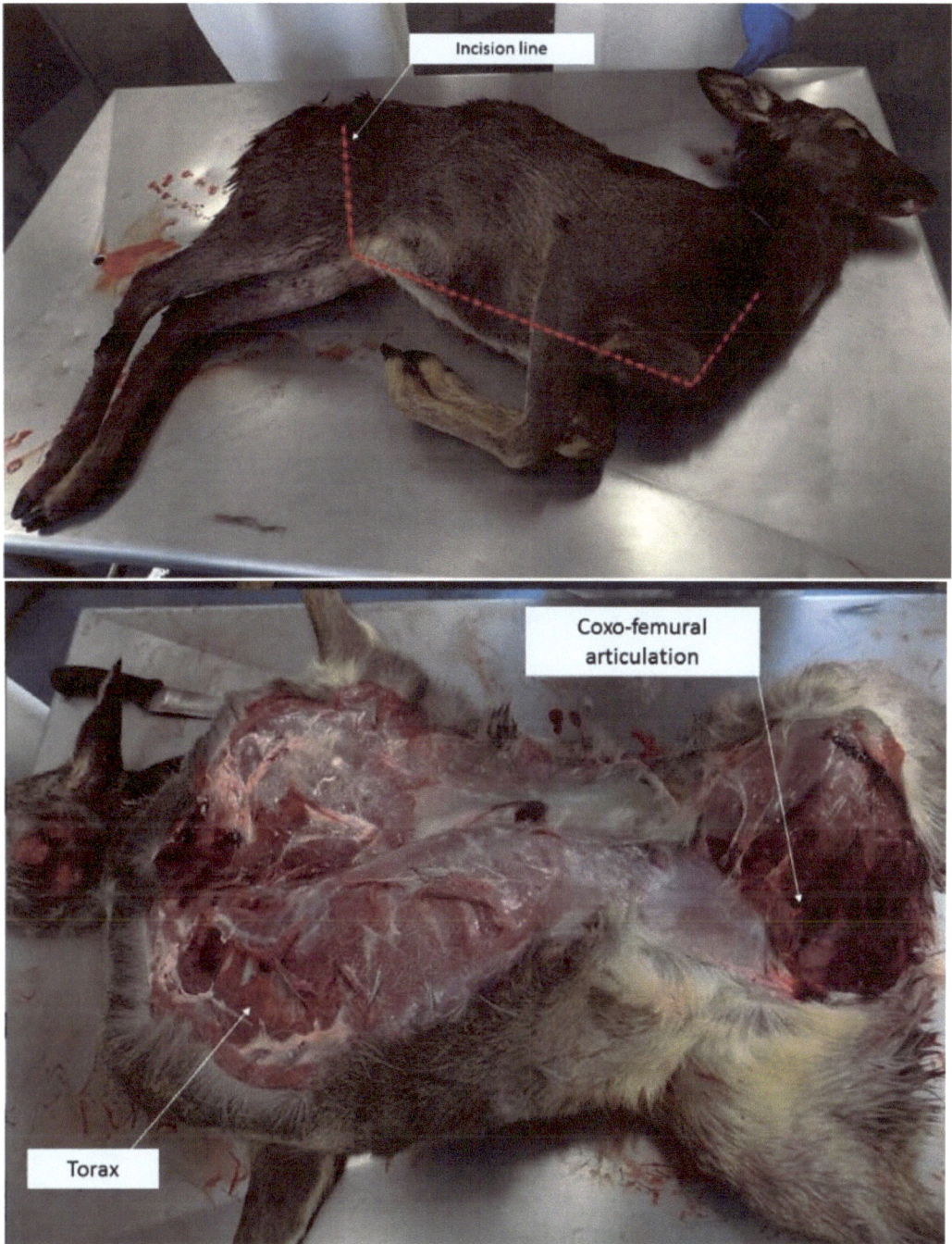

Fig. (20). Exemplification of the incision line and exposition of the pectoral muscles in a *Capreolus capreolus.*

Cardiovascular System

The heart and large vessels should initially be evaluated *in situ*, relative to their position, colour, size and shape. At this stage, in young and new-born animals, we can observe the presence of congenital alterations, such as persistence of the right aortic arch.

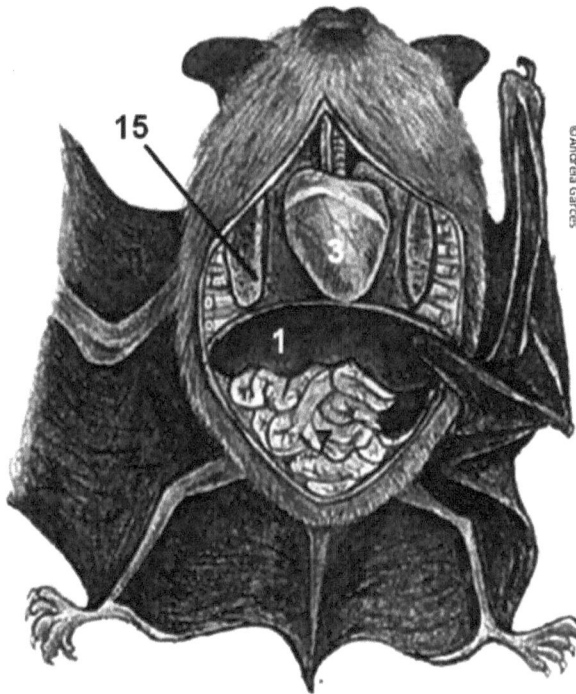

Fig. (21). Representation of the thoracic and abdominal cavity of a *Pipistrellus pipistrellus* (1-liver, 3-heart, 15-lung, 7-intestine).

To have access to the heart, this must be extracted in a block together with the lungs, trachea, thymus and lymph nodes. To do this, one must first remove the tongue and dissect the area of the neck to expose the trachea and oesophagus. The trachea should be incised as cranially as possible and, with some traction, be removed. Then, extracted from the thoracic cavity, we can dissect each organ individually (Fig. **21**).

Extract the heart, trying to leave, as much as possible, the large vessels intact. The palpation of the organ is performed to observe the presence of some alteration. It begins by observing the pericardium externally before making an incision and removing it. When removing the pericardium, it should be noted whether removal is easy or there is adhesion to the surface of the heart. We must observe the fluid present in the pericardium concerning volume, colour and viscosity, and the presence of material such as heart fibrin should be further weighed and measured. It is then dissected, with the use of scissors, following the blood flow. After sectioning the apex, it deforms to expose the two ventricular cavities, open the chambers on the right side and then those on the left side. After exposure to the cardiac chambers, before observing the myocardium and the endocardium, we can observe the blood inside the chambers (Somvanshi and Rao, 2009; King *et al.*, 2013).

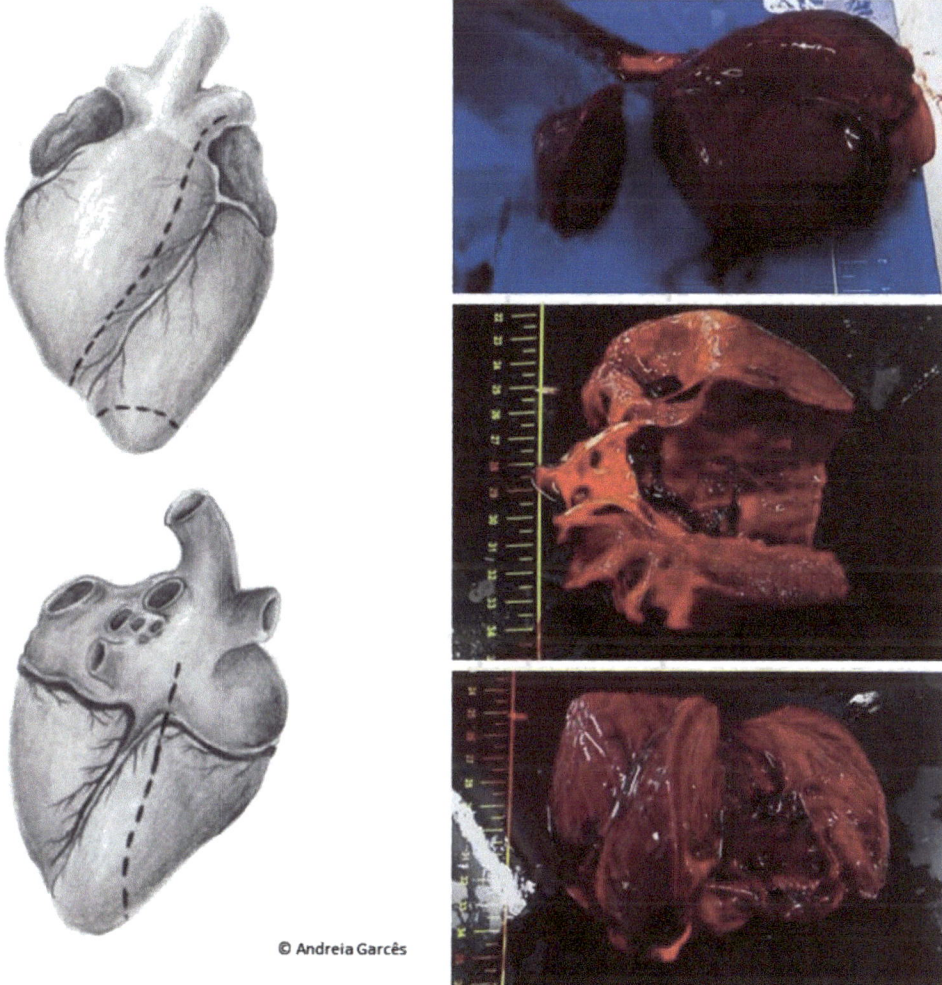

© Andreia Garcês

Fig. (22). Exemplification of the incisions in the myocardium to observe the inside of the atrium, ventriculus, valve, and the apex.

Blood clots should not be observed in the left ventricle. The presence of blood in the left ventricle is suggestive of a recent death, and the examination was performed before rigour Mortis was installed. The presence of clotted blood in the left ventricle suggests that the time of death has already exceeded 24 hours, and the *rigour mortis* phase has passed. However, it may also indicate myocardial disease. Next, observe the heart muscle (myocardium) relative to its thickness and colour. At this stage, it is possible to observe if the ventricles and atriums suffered phenomena of hypertrophy or dilatation. Observe the endocardium and valves to observe the presence of surface irregularities, plaques, thrombi, calcifications, inflammation, among others. Several cross-sections are made in the myocardium to observe the inside of the muscle to detect the presence of non-visible changes in the surface (Figs. **21, 22**) (Munson, 2000; Cardoso, 2002).

Respiratory System

The trachea, bronchi, and lungs are initially evaluated *in situ* for size, location, and staining. The organs are removed by dissecting the tongue, larynx, trachea, and oesophagus inside the jaw (Figs. **23-26**)(Terio, McAloose and Leger, 2018). After applying some traction in the caudal direction, these organs are removed in a block. Palpation is essential, especially in the lungs, because changes in its consistency usually reflect lesions of the same. With scissors, cut along the trachea towards the bronchi to observe the presence of foam, blood, fungal plaques, parasites, foreign bodies or necrotic lesions. Several cuts are also made in the pulmonary parenchyma to observe, for example, the presence of granulomas, parasites, oedema, tumours or abscesses (Fig. **23**) (King *et al.*, 2014).

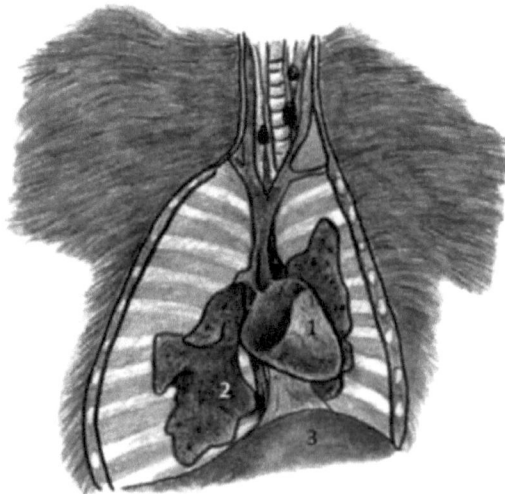

Fig. (23). Representation of fox thoracic cavity (*Vulpes vulpes*) (1-heart, 2-lung, 3-diaphragm).

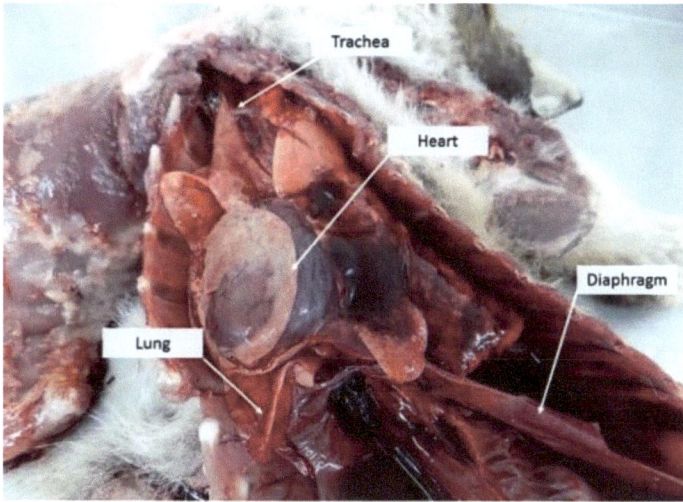

Fig. (24). Representation of the thoracic cavity of a *Vulpes Vulpes*.

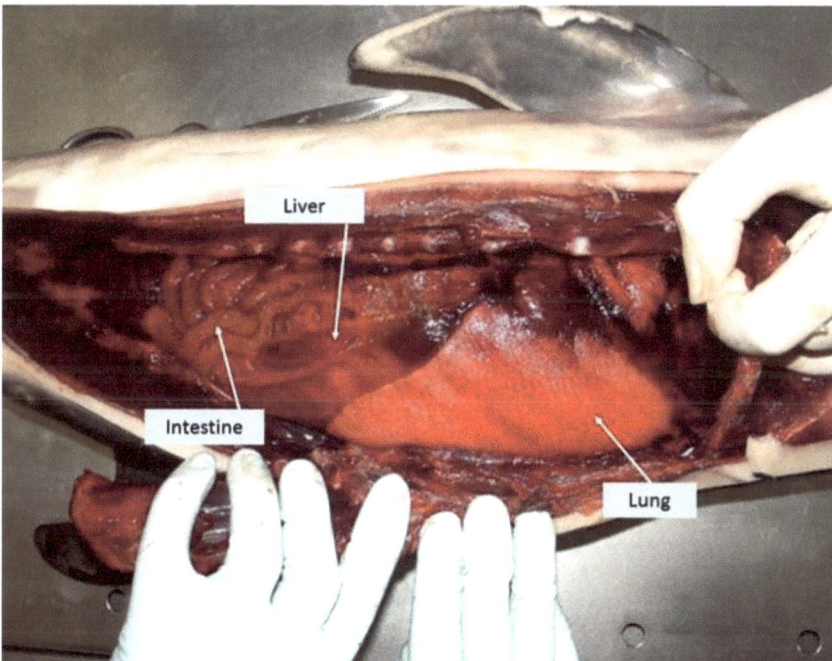

Fig. (25). Thoracic and abdominal cavity in a *Tursiops truncatus*, (photo provided by Daniel Torrão).

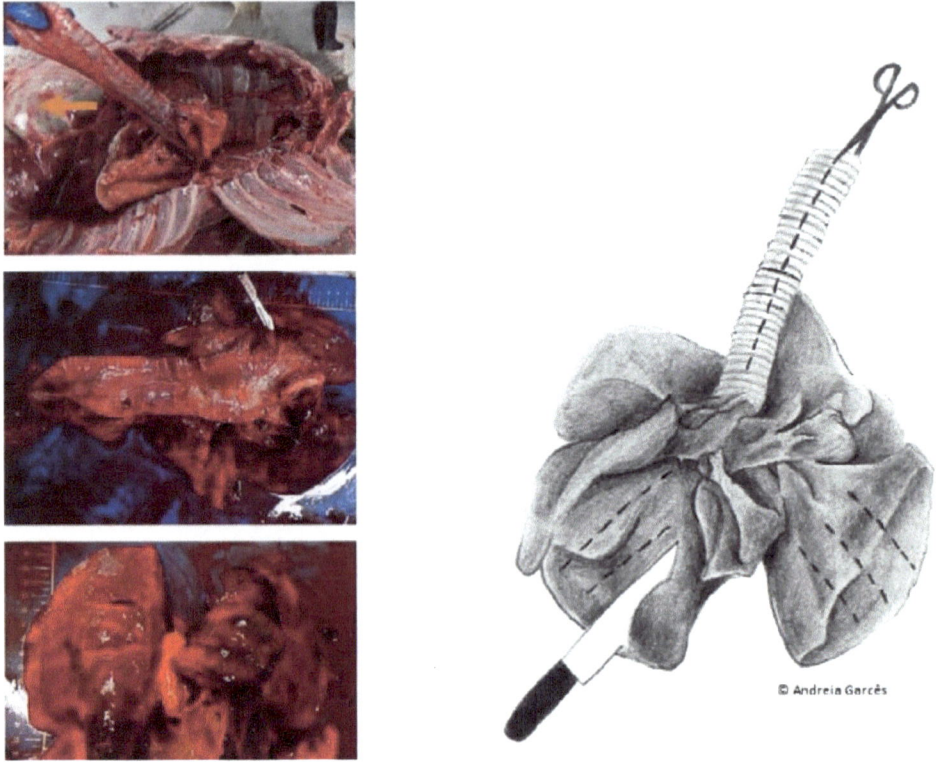

Fig. (26). Exemplification of the traction of the respiratory system and incisions in the trachea and parenchyma to observe the lumen.

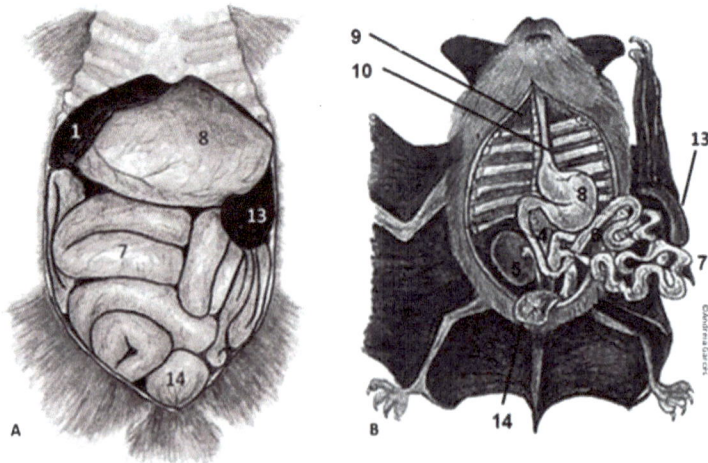

Fig. (27). Representation of the digestive tract in *Vulpes vulpes* **(A)** and *Pipistrellus pipistrellus* **(B)** (1-liver, 3-heart, four pancreas, 5-kidney, 6 - large intestine, 7 - small intestine; 8 – stomach, 9 – trachea, 10 – oesophagus, 13 – spleen, 14 – bladder, 15 – lung, 16 – testis, 17 - adrenal gland).

Digestive System and Annexed Glands

Also, the gastrointestinal tract should be observed *in situ* (Fig. **27**) to detect the presence of strains, volvulus, twists, haemorrhages, discolourations or other injuries. This evaluation should consider the anatomical differences between carnivorous and herbivorous animals. Remove complete treatment (Garcês and Pires, 2017). For this, a possible cranial incision is made in the oesophagus and the most caudal incision in the rectum. These two incisions should be tied with a string so that the contents do not constrict the rest of the organs. With scissors, begin the cranial and focus on the mucosa to expose the lumen (Fig. **28**). The stomach and other gastric compartments should be sectioned by their greater curvature (Fig. **28**) (Cardoso, 2002).

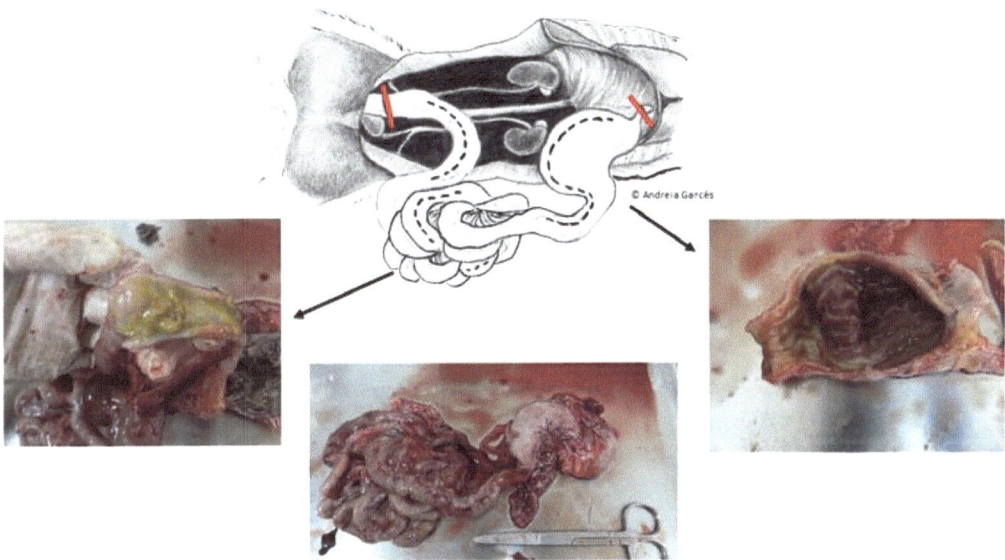

Fig. (28). Exemplification of the incision cuts for removal of the digestive system (red) and line of incision in the stomach and intestine to observe the lumen.

The internal mucosa is observed to detect the presence of alterations such as haemorrhages, ulcers, tumours or other lesions. The contents should be observed and described, as well as the presence of foreign bodies, parasites or other substances. Mesenteric lymph nodes should be observed, palpated and cross-sectioned to observe changes in size, focal lesions, abscesses, metastases or granulomas, for example. In the Equidae family, the abdominal aortic artery, the anterior mesenteric artery, must be observed in detail to detect the presence of thrombi or scars caused by parasites (King *et al.*, 2014).

The liver is examined *in situ*, considering the number of wolves of the species, size and colouration. Observe if there are signs of bleeding, fracture or discolouration areas on the surface (Garcês and Pires, 2017). After removing the liver, methodical palpation and several sagittal cuts are performed to detect the presence of lesions such as haemorrhages, abscesses, parasites or deep discolouration (Fig. **29**). If the gallbladder is present, it should also be observed using the same methodology (Fig. **30-36**) (Kenneth and Rakich, 1994; Cardoso, 2002; McDonough and Southard, 2017).

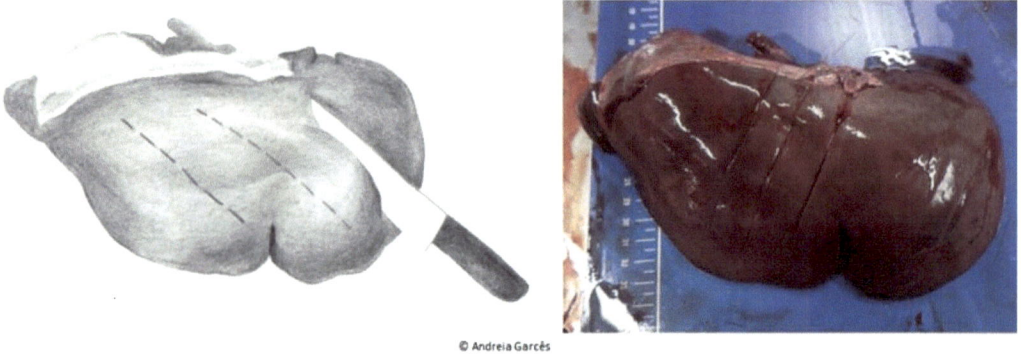

© Andreia Garcês

Fig. (29). Exemplification of the incision cuts in the liver after removal of this.

Fig. (30). Representation of the digestive tract in *Equus zebra* (**1**-heart, **2** - lungs, **3** - liver, **4**-stomach, **5**-spleen, **6**-small intestine, **7**-kidney, **8**-colon, **9**-bladder).

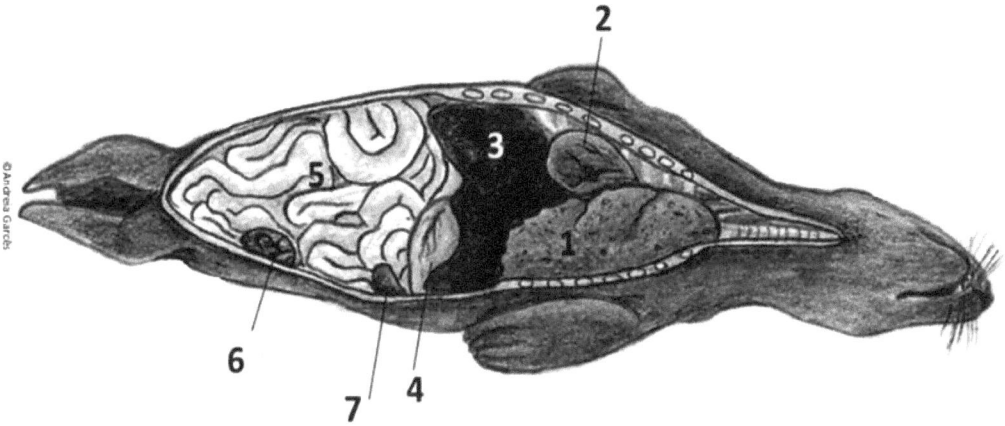

Fig. (31). Representation of the digestive tract in *Hydrurga leptonyx* (**1**-lung, **2**-heart, **3**-liver, **4**-stomach, **5**-intestine, **6**-kidney, **7**-spleen).

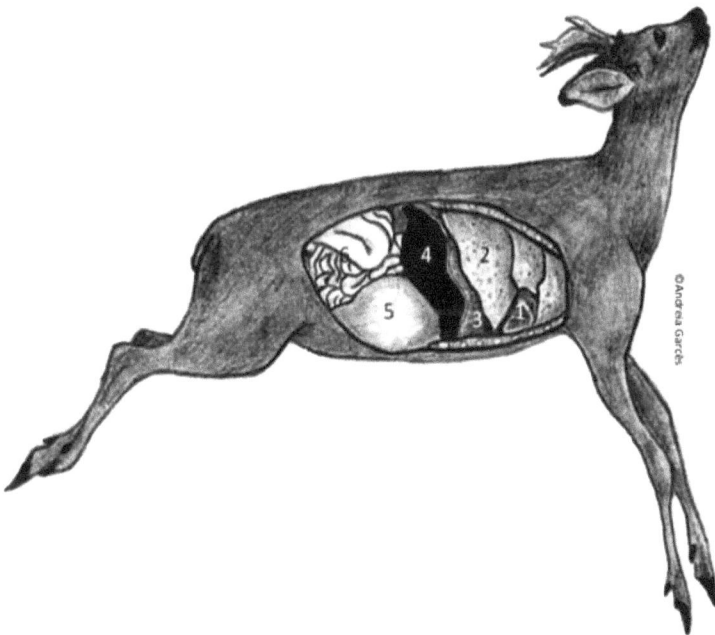

Fig. (32). Representation of the gastrointestinal tract in *Capreolus capreolus* (**1**-heart, **2**-lung, **3**-diaphragm, **4**-liver: **5**-rumen, **6**-intestine).

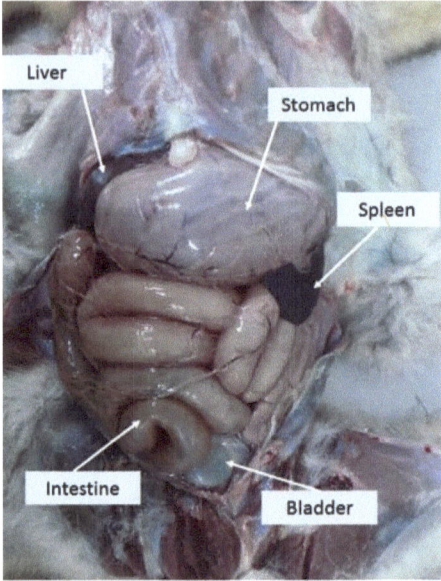

Fig. (33). Representation of the digestive tract in *Vulpes vulpes*.

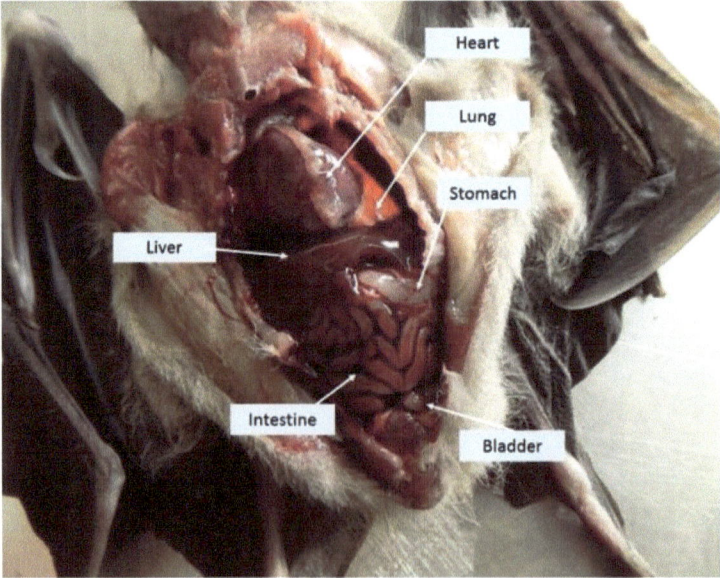

Fig. (34). Thoracic and abdominal cavity in a *Rousettus egyptiacus.*

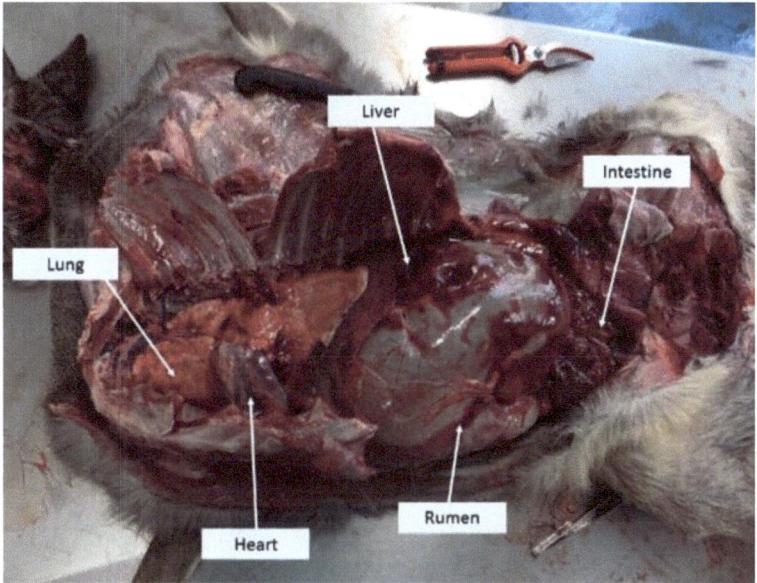

Fig. (35). Representation of the thoracic and abdominal cavity of a *Capreolus capreolus*.

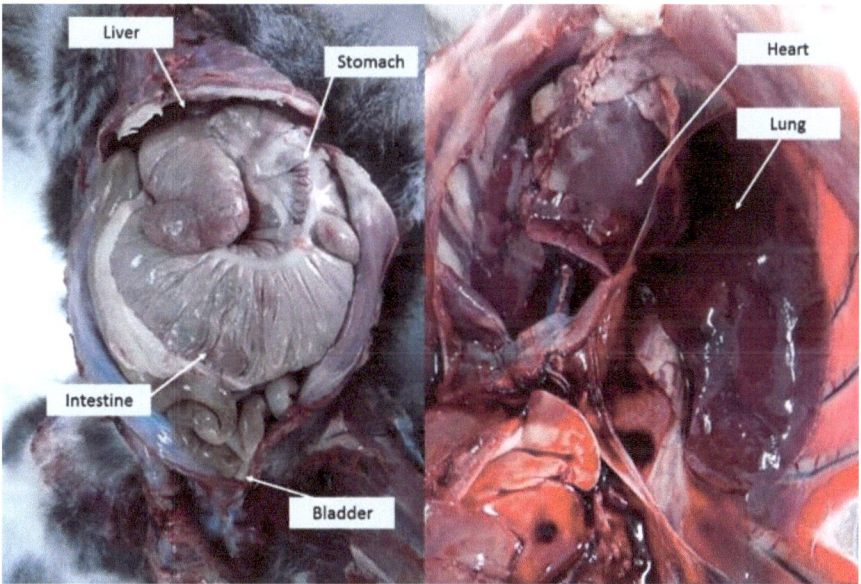

Fig. (36). Representation of the thoracic and abdominal cavity of a *Macropus rufagriseus.*

Endocrine and Hematopoietic System

The spleen should be observed *in situ* to observe the location, shape appropriate to the species, size and colouration. If the spleen is black and has an earthy appearance, it should be tested for Bacillus anthracis before proceeding (Munson,

2000). After removing the spleen, methodical palpation and section of this organ should be performed to observe the presence of lesions, such as haemorrhages, abscesses, swelling, areas of discolouration,*etc*. Splenomegaly can be due to euthanasia with barbiturates (Fig. **37**) (King *et al.*, 2014).

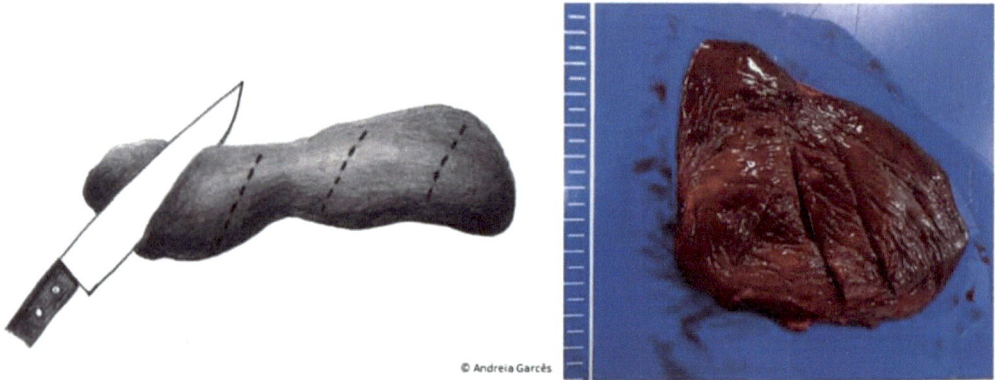

Fig. (37). Exemplification of the incision cuts in the spleen after removal of this.

Urinary System

The kidneys should be observed *in situ*, taking into account the anatomical peculiarities of each species, and their size. We must observe the shape, the colouration and the anatomical location (Figs. **38 & 39**) (Cardoso, 2002; McDonough and Southard, 2017).

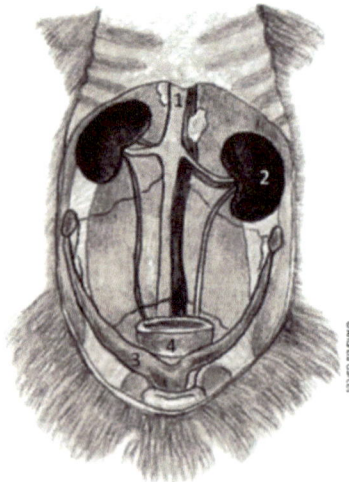

Fig. (38). Representation of the genital and urinary tract in *Vulpes vulpes* (1-adrenal gland, 2-kidney, 3-uterus, 4-intestine).

The ureters are observed and palpated for calculus detection, and unusual changes (*e.g.* , abnormal developments or ectopic ureters). The kidneys are removed, and the perirenal fat is evaluated. The palpation of the kidney is performed, and it is separated, then decapsulating, considering if there are adhesions or alterations of the renal surface.

The cortex and the medulla and pelvis are then preserved if there is any differentiation between them, cortex/medulla, or other lesions such as cysts, parasites, or haemorrhages. The bladder is dissected to observe the mucosa, the presence of abnormal contents, or other lesions (Fig. **40**) (King *et al.*, 2013).

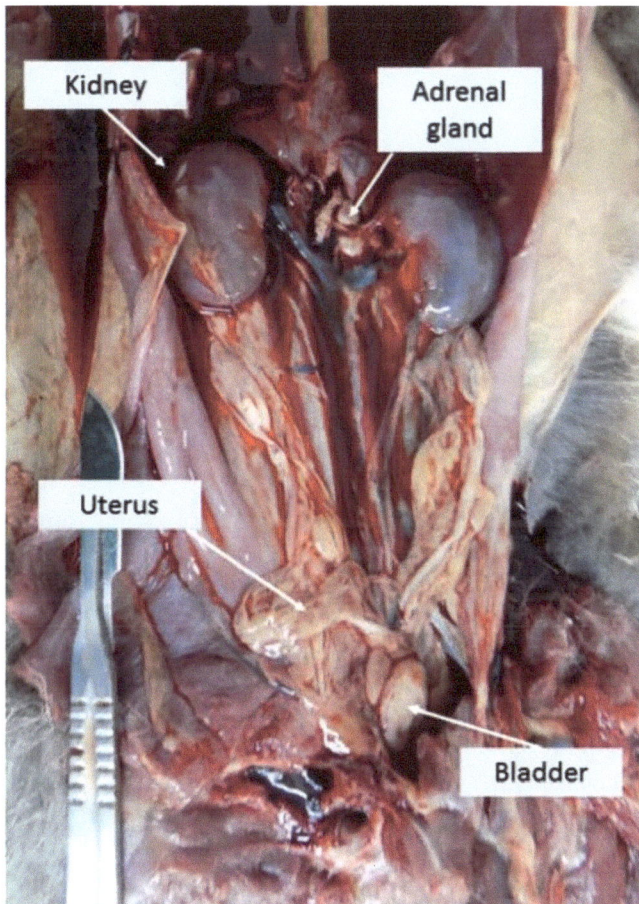

Fig. (39). Representation of the genital and urinary tract in *Vulpes vulpes.*

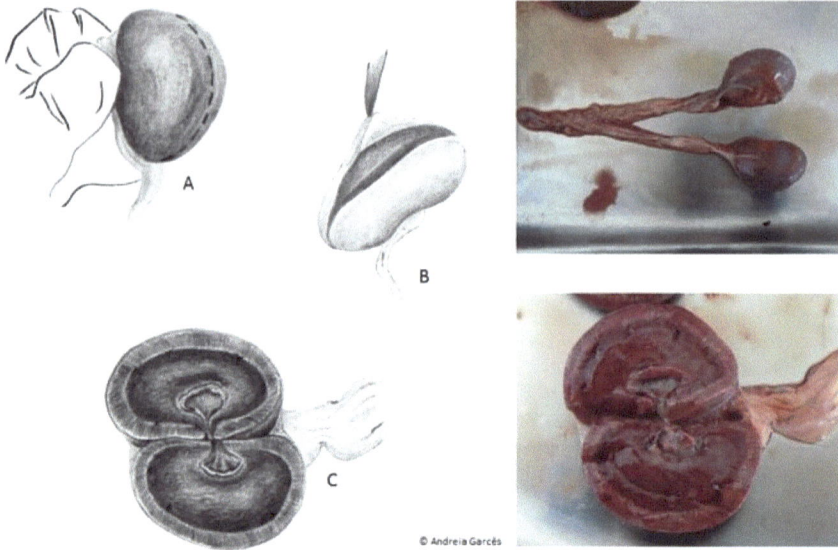

Fig. (40). Exemplification of the steps of incision cuts in the liver in the kidney, including decapsulation in B.

Genital System

The genital organs are observed *in situ* to detect changes in colour or size. They should, after removal, be methodically palpated and sectioned. The mucosa and the presence of abnormal contents such as purulent material or blood are observed in the uterus and uterine horns (Garcês and Pires, 2017). It is also observed if there are foetuses or changes in the placenta (Fig. **41**). The same procedure should be performed for the organs of the male genitalia (King *et al.*, 2013).

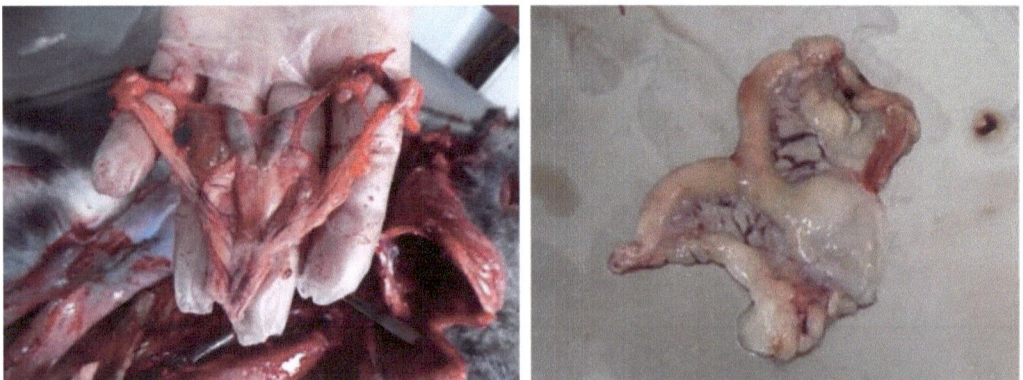

Fig. (41). Female genital apparatus in different species.

Nervous System

The brain is one of the first organs to enter into autolysis, and therefore, it is not often possible to conclude its examination.

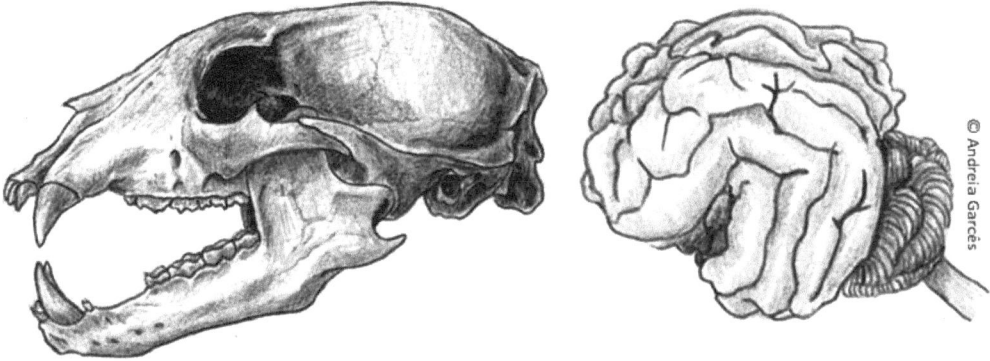

Fig. (42). Representation of the skull and brain of *Ursus arctos*.

To access this organ, begin by removing the skin from the skull and then the muscles. After the bone is exposed with a bone saw, a transverse incision is made between the orbits, two cuts beginning at the ends of the cut performed previously toward the foramen magnum.

After completing the cuts, raise the top of the skull to expose the intracranial cavity. It is observed the brain and the meninges registering the changes as the presence of haemorrhages, abscesses, inflammations, parasites, among others (Figs. **42** & **43**) (King *et al.*, 2014; Zachary, 2016; Terio, McAloose and Leger, 2018).

Musculoskeletal System

The skeleton has already been partially evaluated by palpation of limbs. During the opening of the corpse, we should also observe some injuries such as bleeding in the bones, fissures, bruises, fractures of bones. We must also section the joints to see if they present any changes such as inflammations, fractures or foreign bodies (King *et al.*, 2014).

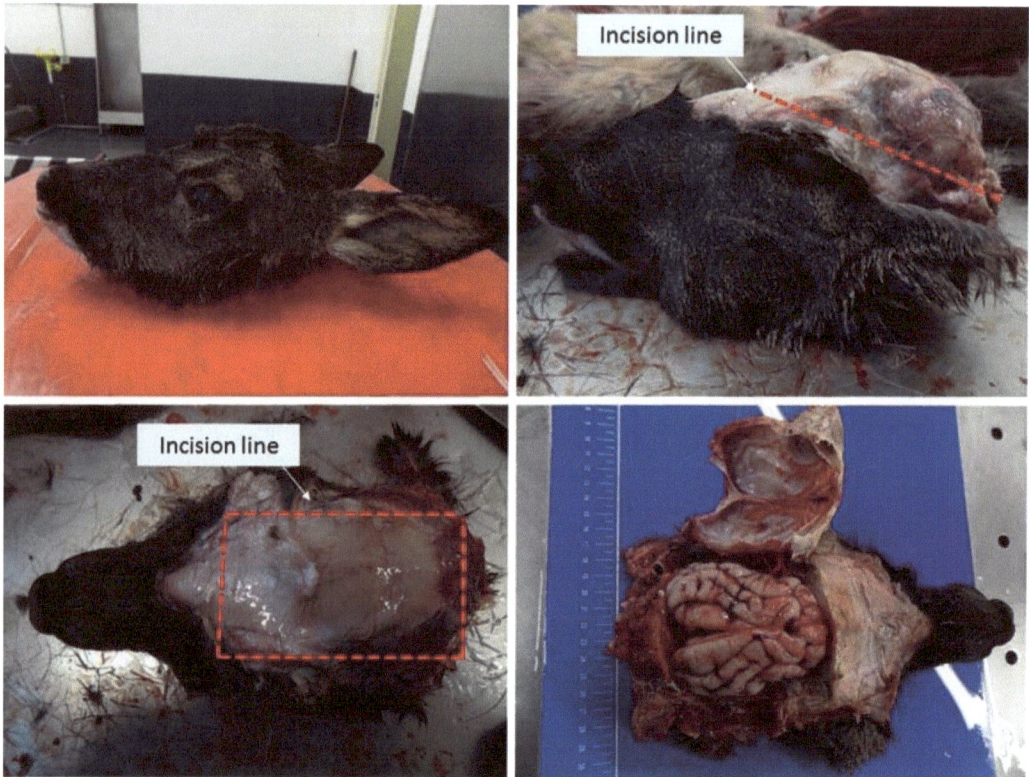

Fig. (43). Open skull and brain of *a Capreolus capreolus*.

REFERENCES

Cardoso, C (2002) Técnica de necropsia.*Animais de Laboratório: criação e experimentação* Fiocruz, Rio de Janeiro 332-5.

Cramer, S, Ketten, DR & Montie, EW (2007) *Odontocete Salvage, Necropsy, Ear Extraction, and Imaging Protocols.* University of North Carolina at Wilmington, Wilmington 171.

Eros, C (2007) *Procedures for the Salvage and Necropsy of the Dugong (Dugong dugon).* Great Barrier Reef Marine Park Authority, Australia 104.

Friend, M & Franson, JC (1999) *Field Manual of Wildlife Diseases - General Field Procedures and Diseases of Birds.* Library of Congress, Cataloging, USA.

Garcês, A & Pires, I (2017) *Manual de Técnicas de Necrópsia em Animais Selvagens* Arteology, Porto.

Kenneth, L & Rakich, P (1994) NECROPSY EXAMINATION.*Avian Medicine: Principles and Application* Wingers Publishing, Florida 355-78.

King, JM (2013) *The necropsy book: A Guide for Veterinary Students, Residents, Clinicians, Pathologists, and Biological Researchers.* College of Veterinary Medicine, Cornell University, NY, USA.

King, JM (2014) *The necropsy book: A Guide for Veterinary Students, by The Necropsy Book.* Charles Louis David DVM Foundation Publisher, Ithaca.

McDonough, SP & Southard, T (2017) *Necropsy Guide for Dogs, Cats, and Small Mammals.*Wiley-Blackwell, Ames, Iowa.

[http://dx.doi.org/10.1002/9781119317005]

Mclellan, W, Rommel, S & Pabst, DA. *Right whale necropsy protocol.* Marine Mammal Health and Stranding Response Program, Silver Spring.

Munson, L (2000) *Necropsy of Wild Animals.* Wildlife Health Center, School of Veterinary Medicine, USA.

Somvanshi, R & Rao, JR (2009) *Necropsy techniques and necropsy conference manual.* Veterinary Research Institute, India.

Raverty, SA, Gaydos, JK & St. Leger, J (2014) *Killer whale necropsy and disease testing protocol.* International Association for Aquatic Animal Medicine.

Terio, KA, McAloose, D, St. Leger, J (2018). *Pathology of Wildlife and Zoo Animals.* Academic Press, UK.

Wobeser, G (2006). *Essentials of Disease in Wild Animals.* Blackwell Publishing Ltd., Iowa, USA.

Woodford, MH, Keet, DF & Begins, RG (2000) *Post-mortem procedures for wildlife veterinarians and field biologists.* Iucn, Paris, France.

Zachary, JF (2016) *Pathologic basis of veterinary disease.* 6th. Mosby, USA.

CHAPTER 6

Necropsy in Reptiles

Outline: In this chapter, we describe the method of necropsy in wild reptiles, offering some information regarding the different orders and anatomic characteristics of determined species.

Keywords: Animals sentinels, Conservation, Crocodiles, Lizards, Mortality, Necropsy, Pathology, *Post-Mortem*, Snakes, Turtles, Wild Reptiles.

GENERAL CONSIDERATIONS

The Reptilia class has a large diversity of species with different anatomical features. The *post mortem* examination technique has some variations depending on the group approached. In this chapter, three orders/suborders were considered: Order Testudinata, Suborder Sauria and Suborder Ophidia (Fig. **1**) (Divers and Mader, 2005).

Fig. (1). Representation of reptiles of different orders (Chelonia, Squamata, crocodile).

Reptiles, both in freedom and captivity, are usually found in hot and humid habitats, due to their biologic characteristics of cold-blooded animals, both in terrestrial and aquatic environments. The decomposition process of their carcass is accelerated not only due to the characteristic of their habitat but also due to their rapid metabolism, especially in smaller animals (Jacobson, 2007). Preferably, the dead animal should be collected and refrigerated (about 4-10° C) as soon as possible.

In larger animals, it may be useful to dip them in ice water or make an incision in the abdomen to accelerate the cooling process. Freezing should always be used as a last resort (Somvanshi and Rao, 2009).

An important issue to take into account with these animals is the perception of their death. Due to the anatomical characteristics of the chelonians and the presence of bone scales or plaques, it is sometimes impossible to auscultate the heart. Often these animals go into hibernation and are mistakenly considered dead (Jacobson, 2007). In animals that have been euthanized, especially in the case of the chelonians, many authors advise performing their decapitation before performing the necropsy to make sure that they are dead. However, it should be noted that, even after death, reptiles often have involuntary movements of the skeletal and cardiac muscles (Survey, 2000; Mader, 2006; Flint *et al.*, 2009).

Reptiles undergo skin changes, a process known as moulting. It is their way to get rid of the old skin and gain new space to grow. The snakes lose their skin completely, while the chelonians, lizards and crocodiles lose it in fragments or plaques (O'Malley, 2005). Therefore, during the external examination of these animals, the presence of dry and elevated skin and opaque eyes may be indicative that the animal was in moult. It should be assessed whether the moult was occurring correctly or if there were any pathological changes (Divers and Mader, 2005).

SUBORDER OPHIDIA

General Considerations Before the Necropsy

About 2900 species of snakes exist worldwide. These can be grouped into four main families: Boidae (constrictors, most primitive), Colibridae, Elapidae (more evolved with some very poisonous species) and Viperidae (poisonous) (Jacobson, 2007).

Their body is elongated with a more or less cylindrical shape and the abdomen flattened for easier locomotion. The skeleton consists of 400 vertebrae, each with its ribs and skeletal muscles. They do not have a sternum and the first two vertebrae do not have associated ribs. Their size is variable. *Eunectes Murinus* can reach 10m, while *Typhlopidae* species range between 10 and 30cm (O'Malley, 2005).

External Exam

Before the necropsy begins, it is necessary to identify the species. There are several venomous species within this class, and therefore some care is necessary before starting the *post mortem* examination. In these animals, the jaws should be kept closed using adhesive tape or by placing the head in a cylindrical container (for example a syringe, depending on the size of the animal) attached with an adhesive. Then the head should be decapitated and placed in a container with formalin to deactivate the venom before performing the head examination (Jacobson, 2007). However, as there is no certainty that all the constituents of the poison will be deactivated, it is still necessary to be very careful when handling the head, especially the poisonous glands and the poison inoculating teeth. Also, the use of eye protection is advisable, such as glasses or a mask (Divers and Mader, 2005).

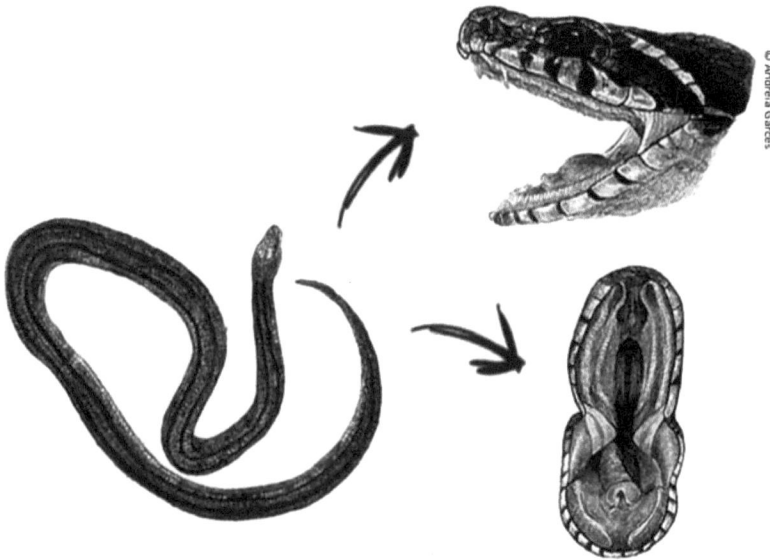

Fig. (2). Representation of the external examination of a snake.

The skin should then be examined for lesions suggestive of trauma, scars, changes in scales or ectoparasites. Skin with or without lesions should be collected and fixed for histopathological examination (King *et al.*, 2014).

Palpation of the animal is also important, allowing the detection of fractures, dislocations, masses, oedema or other lesions.

The dentition differs with the species, but most species have a row of teeth in each lower jaw and two in each maxillary and palatine or pterygoid bone of the upper jaw. The teeth are long and curved into the oral cavity since the only function is to fix the prey and not to chew it. They are pleurodont. Some poisonous species have some of the maxillary teeth modified for the inoculation of the venom (Jacobson,

2007). They should be counted and compared with the dental formulas. Injuries such as broken teeth, fistulas, abscesses, masses or other changes should be observed (King *et al.*, 2013; Garcês and Pires, 2017; Brooks, 2018).

The venom gland normally lies in the caudal region to the eye with some changes depending on the species. The gums should also be seen in the mouth to detect signs of stomatitis or gingivitis (Divers and Mader, 2005).

The tongue has a function of smell, taste and touch. It is elongated, thin and forks at the end.

Fig. (3). Representation of the external examination of a *Pantherophis guttatus.*

Other natural orifices such as eyes, nostrils, ears, and cloaca should also be observed to identify possible changes. The eyes have no eyelids or nictitating membrane, possessing a protective spectacle of the cornea. In addition to the nostrils, the Jacobson's organ is positioned in the upper jaw (Jacobson, 2007). Some species (Boas and Pythons) have small holes in varying numbers in the upper and lower lip scales, which are infrared detectors (Figs. **2-5**) (Girling and Raiti, 2004; Terio, Macloose and Leger, 2018).

The body condition is evaluated as described below for the animals of the Order Sauria.

Internal Exam

The carcass is placed in a ventrodorsal position. To facilitate the localization of the organs, the body can be divided into three parts: cranial region (heart, trachea, oesophagus, thyroid, parathyroid, thymus, proximal lung), medial region (stomach, liver, lung, spleen, and pancreas) and region caudal (small and large intestine, kidneys and gonads)(Girling and Raiti, 2004; King *et al.*, 2014; Terio, Macloose and Leger, 2018).

Fig. (4). Representation of the external examination of a snake, in position ventrodorsal (**A**) and dorsoventral (**B**) position.

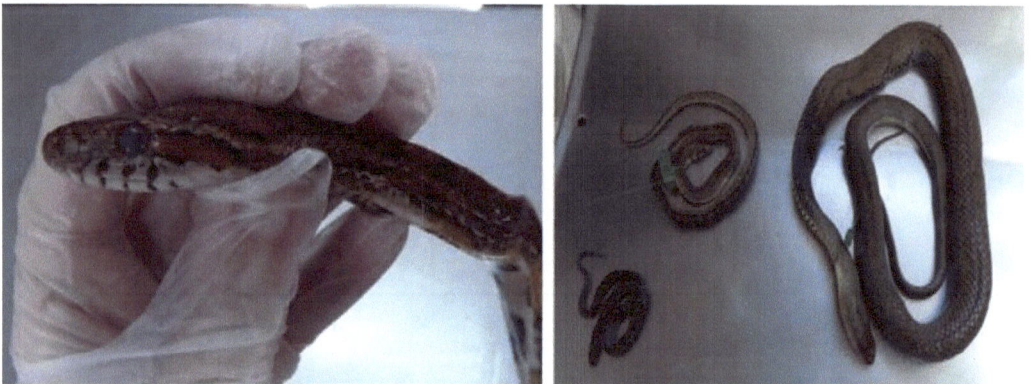

Fig. 5 cont.....

Fig. (5). Representation of different species belonging to the suborder Ophidia.

Then, with scissors or a scalpel, an incision is made from the cloaca to the intermandibular space following the midline (Fig. **6**). It should not be performed directly on the midline, but a little further to the left or right due to the ventral abdominal vein. In larger species, the ventral scales are larger and harder so it is necessary to perform the first incision laterally (Mullineaux, Best and Cooper, 2003; Divers and Mader, 2005).

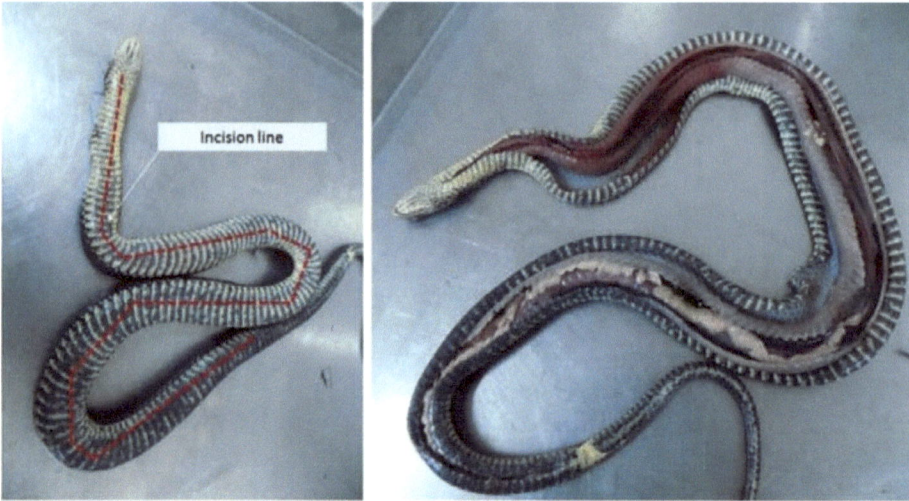

Fig. (6). Representation of the incision line in a snake.

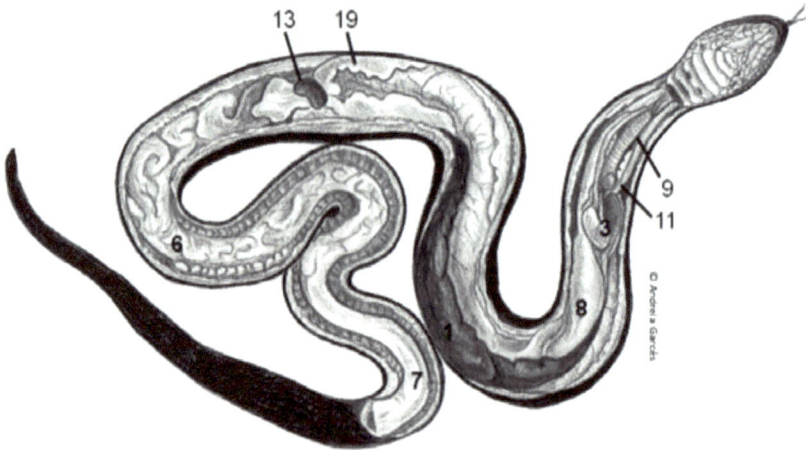

Fig. (7). Scheme of the coelomic cavity after the first incision and removal of the skin (**1** -liver, **3**-heart, **6**-large intestine, **7**-small intestine, **8**-stomach, **9**-trachea, **11**-thyroid, **19**-fat).

After this incision has been made, the skin is rebounded laterally to expose the muscle. The intermandibular joint should then be incised to facilitate mouth inspection (Fig. **7**). At this time, skin and muscle samples should be collected (Jacobson, 2007). When making the incision in the midline, we have access to the coelomic cavity. In its caudal portion, it is possible to observe the prominent fat bodies, that give us indications about the nutritional state of the animal. They should be removed to allow a better observation of the viscera in the coelomic cavity (Fig. **8**) (Munson, 2000).

In the cranial portion to the heart, we can see the thymus, the parathyroid and the thyroid. The thymus is an organ that has two lobes, one cranial and one caudal. The thyroid is a singular structure (Jacobson, 2007). These animals have two parathyroids, with one of the pairs located between the lobes of the thymus and the other located at the bifurcation of the carotid artery (Divers and Mader, 2005).

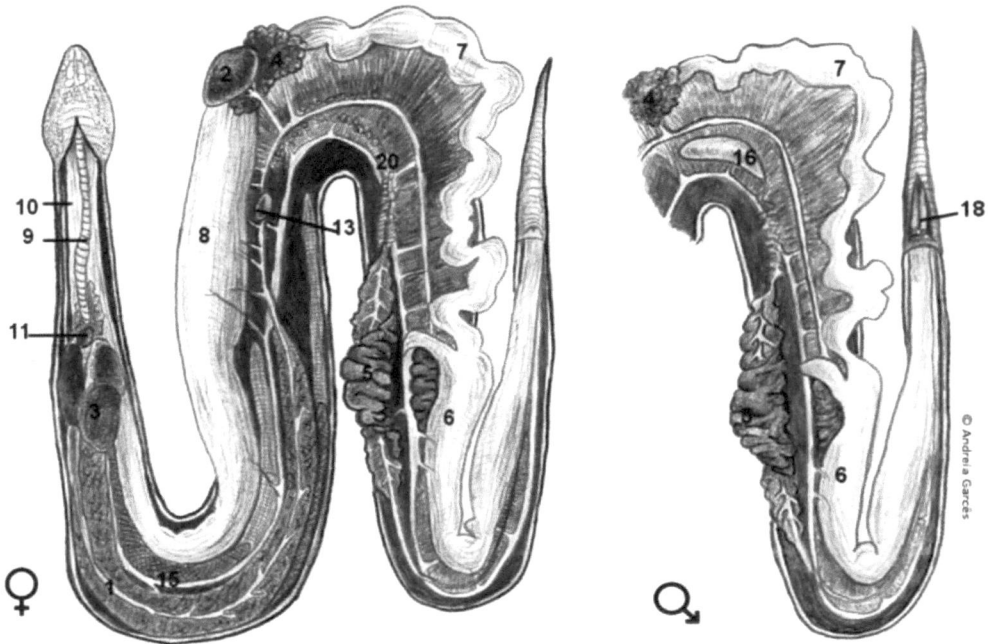

Fig. (8). Representation of the internal organs of a female snake (**1**-liver, **2**-gallbladder, **3**-heart, **4**-pancreas, **5**-kidney, **6**-large intestine, **7**-small intestine, **8**-stomach, **9**-trachea, **10** - oesophagus, **11** - thyroid, **12** - parathyroid, **13** - spleen, - **14** - bladder, **15** - lung, **16** - tests, **18** – hemipenis; **20** - ovary).

The organs of the coelomic cavity, after having been observed *in situ*, can be removed in a single block. The trachea and oesophagus should be sectioned caudally to the pharynx. Then with tweezers, traction is performed caudally, while at the same time a blunt dissection is performed, or the connective tissues sectioned with a scalpel or scissors. Finally, an incision is made next to the cloaca. The viscera removed in a block should be placed on the work table for examination (Munson, 2000; Terio, Macloose and Leger, 2018).

The heart has three chambers (two auricles and one ventricle) and is elongated and thin. It is located in a cranioventral location at the bifurcation of the trachea.

The pericardial sac should be sectioned to observe the presence of fluid, which in reduced amounts is physiological. Its colour and presence of substances such as blood or fibrin should be observed (Jacobson, 2007). The heart should be palpated and visually examined for its shape, colour, trauma, masses or foreign bodies. To observe the valves, myocardium and endocardium, chambers are opened. Regarding the great vessels, the left and right aorta merge to form the caudal dorsal aorta to the heart (Jacobson, 2007). They should be inspected with particular attention to detect changes such as arteriosclerosis or dystrophic mineralization. More developed snakes like the colubrid and vipers have only the left carotid artery well developed while the right is a rudimentary structure. However, in other families, they are relatively symmetrical (Girling and Raiti, 2004; Divers and Mader, 2005).

The respiratory system can then observed. Examine the pharynx, trachea, and bronchi. The trachea has incomplete cartilaginous rings. An incision is made along the trachea with the help of scissors to observe the presence of alterations in the mucosa and presence of abnormal contents (liquid, food, foreign bodies, parasites) in the lumen (Jacobson, 2007).

In snakes, the two lungs may have different sizes, depending on the family. The left lung is smaller or completely absent in some species. In the family of Vipers, they only have the float function. In the Colunbrids, only one lung is functional, and the left lung is vestigial. Boias have both lungs developed. The lungs should be palpated carefully to detect the presence of nodules or other lesions. Then, cross-sections should be performed to evaluate the presence of parasites or exudates. The right lung extends from the heart to the cranial portion of the right kidney. The aquatic snakes have an air sac that extends caudally toward the cloaca and should be observed to detect the presence of foreign bodies or masses (Jacobson, 2007). It is a very simple system, and it is sometimes difficult to differentiate the various parts during the *post mortem* examination. Therefore, several samples should be removed from different areas for later histopathological examination (Girling and Raiti, 2004; Divers and Mader, 2005).

The oesophagus, stomach, small intestine and large intestine should be first examined visually. The cranial portion of the oesophagus has goblet cells in the mucosa and submucosal lymphoid tissue associated with the gastrointestinal tract (GALT). In the Boias, it is organized in oesophageal tonsils. Using a pair of scissors, they should be opened and the internal mucosa in the lumen should be examined for possible lesions such as gastric ulcers, trauma, masses, or other injuries (Girling and Raiti, 2004; Divers and Mader, 2005).

The mucosa of the small intestine in these animals is very thick and muscular which should not be confused with an abnormality. Care should be taken of its contents, due to the possibility of being presence chemicals, parasites or foreign bodies. Gastrointestinal contents may be collected whenever poisoning is suspected (Woodford, Keet and Begins, 2000).

The cloaca is linear and divided into three sections. Some species have glands in the cloaca that produce fetid secretions used to repel predators (Jacobson, 2007).

The pancreas is a smooth or multilobate organ, pale in colour, located caudal to the spleen and near the duodenum. Some species have the pancreas fused to the spleen, called spleen pancreas. This should be observed as fresh as possible as it degrades rapidly (Jacobson, 2007).

The liver is elongated, brown. In most species, the gallbladder lies distally to it, and it is connected by a long bile duct. These should be examined visually and palpated to identify abnormal structures. In the liver, cuts should be made to detect, in-depth, the presence of parasites, areas of necrosis, abscesses or alteration of the walls of the canaliculi and blood vessels. Section of the gallbladder should be performed to observe the contents and check for obstructions, calculations or mucosal changes in the lumen (Woodford, Keet and Begins, 2000).

The kidneys are multilobed, elongated and located cranially to the cloaca. They have a dark brown colour but during the reproductive activity of the males, they may be paler due to the formation of sexual segments. The kidneys should be examined for the presence of urates, chronic kidney disease, urolithiasis, bacterial infections, parasites or neoplasms. These animals do not have a urine bladder (Fig. **9**) (Meredith and Redro, 2002).

The adrenal glands are thin, elongated structures with a yellowish colour. They are inserted in the connective tissue that supports the gonads (mesorquium and mesovarium) (Jacobson, 2007).

The testes have an intracellular location, located between the pancreas and the kidneys. The males have two hemipenes, which are located in the caudal extension of the cloaca at the base of the tail (Jacobson, 2007). The ovaries are pairs and are located asymmetrically near the pancreas. The right ovary is usually larger than the left and located more cranially (Girling and Raiti, 2004).

Snakes may be oviparous or viviparous. The right ovary is prolonged in the oviduct in the case of the oviparous or the oviduct, uterus and vagina in the viviparous. In snakes of the genus *Typhlops* and *Leptotyplops,* the left oviduct does not exist (Jacobson, 2007).

The gonads organs should be removed and observe if there are changes such as bacterial infections, pre or post-ovulatory stasis, prolapses of the uterus/hemipenis, mineralization or impaction of the hemipenis, nodules or abscesses (Girling and Raiti, 2004; Divers and Mader, 2005).

To observe the brain, start with the decapitation of the animal at the level of the atlas. In small animals (skull less than 2 cm), the head can be placed directly into the formalin and then decalcified to section the brain. In larger animals, an incision is made on the skin, and it is removed to expose the skull (Jacobson, 2007). Then, depending on the size of the animal, the skull can be opened with a pair of scissors or a bone saw). Likewise, cross-sections can be performed on the spine to observe the bone marrow and spinal canal and withdraw samples if necessary (Girling and Raiti, 2004; Divers and Mader, 2005).

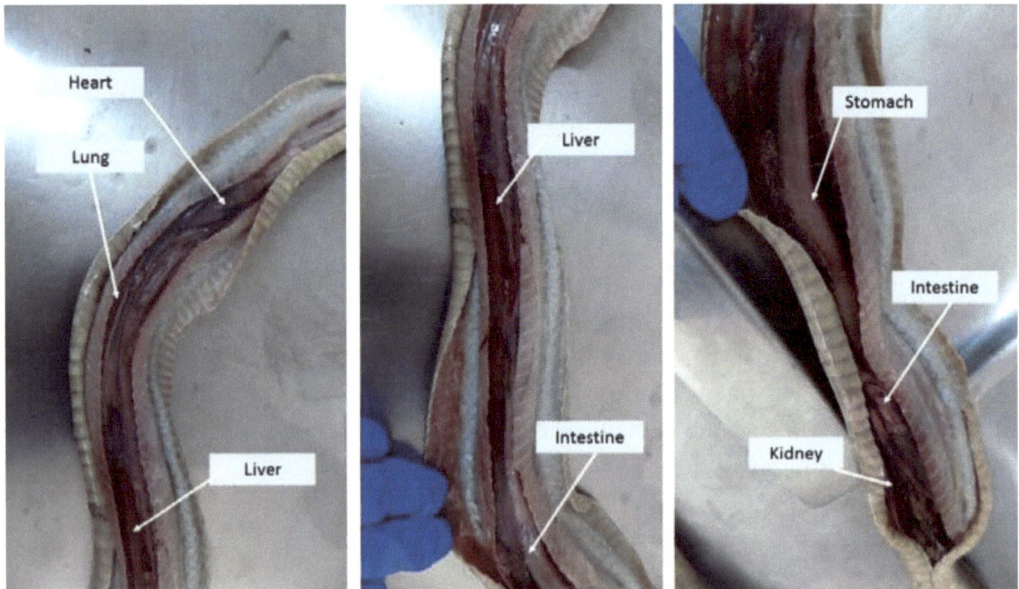

Fig. (9). Internal exam in a *Rhinechis scalaris.*

SUBORDER SAURIA

General Considerations Before the Necropsy

The size of these animals is very variable with the species, ranging *from Varanus komodiensis* with 3 meters to *Spaerodactylos sp.* with only 2 cm. This group inhabits a wide variety of habitats, all over the world, except for the continent of Antarctica. There are worldwide around 5500 species (Fig. **10**). Some lizards of the Scincic family do not have limbs resembling snakes. There are some species such as *Heloderma suspectem, Varanus komodiensis* and *Heloderma horridum* which produce neurotoxins through sublingual glands (Jacobson, 2007). In this case, the pathologist must take the necessary safety measures when handling poisonous species, the method here being the same as described for venomous snakes. (Girling and Raiti, 2004; Terio, Macloose and Leger, 2018).

External Exam

The examination begins with the identification of the species. The skin should be examined for possible traumatic lesions, scars, changes in pigmentation or scaling symmetries, ulcers, Pox Virus lesions, abrasions (more on the cheek area and the plantar area of the limbs) ectoparasites or other lesions. The carcass is palpated to detect fractures, dislocations, masses, oedema or other lesions. In the limbs, in addition to these described lesions, we can also observe amputations, arthritis, lack of digits or pododermatitis (Divers and Mader, 2005).

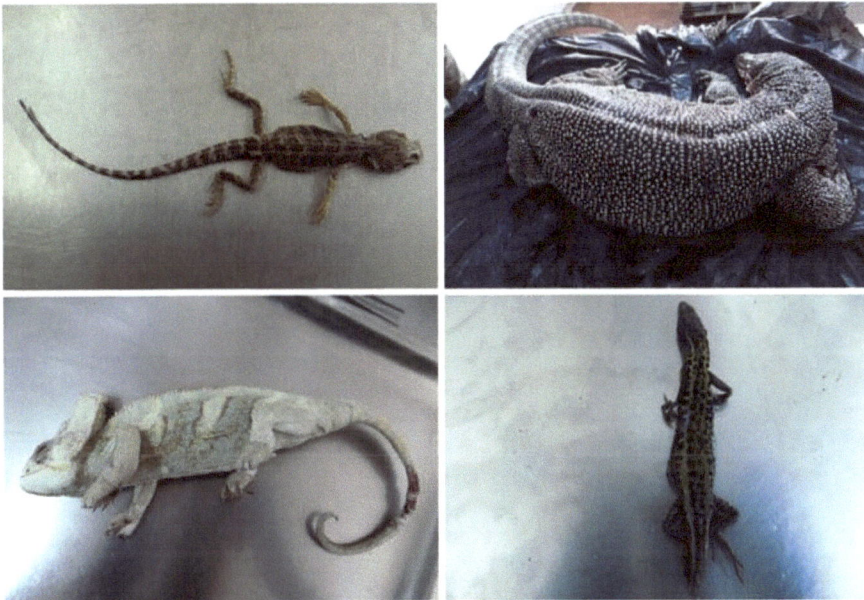

Fig. (10). Representation of different species from the Suborder Sauria.

These animals have two types of dentition: the acrodes (Agamidas, Chameleons) or pleurodont (Iguanas). The first type, whenever a tooth is broken or lost it is regenerated, while in the other it is not (Jacobson, 2007). They should be counted and compared with dental formulas, and investigate any lesions such as broken teeth, fistulas, abscesses, masses or others. In the mouth also observe the gums (Fig. **11**) (Woodford, Keet and Begins, 2000).

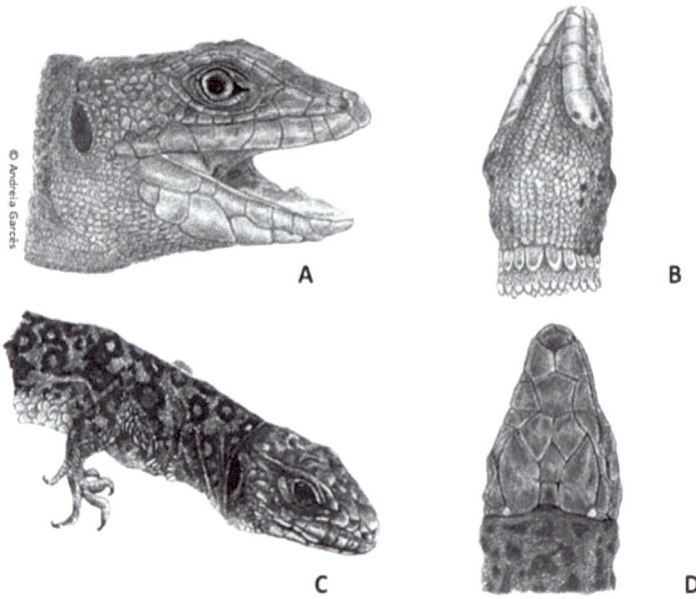

Fig. (11). Detail of the head in a lizard with prominence to the oral cavity **(A)**, ventral view **(B)** and dorsal view **(D)**.

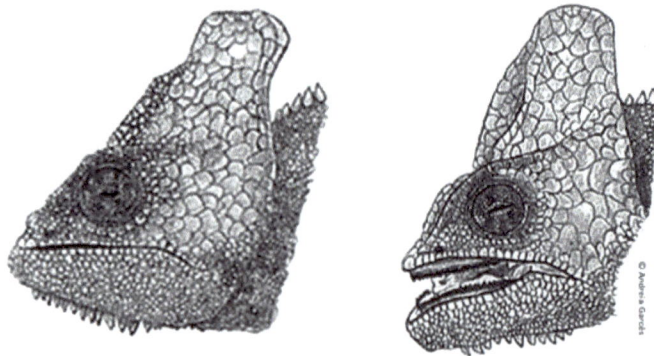

Fig. (12). Detail on the head and mouth of a chameleon.

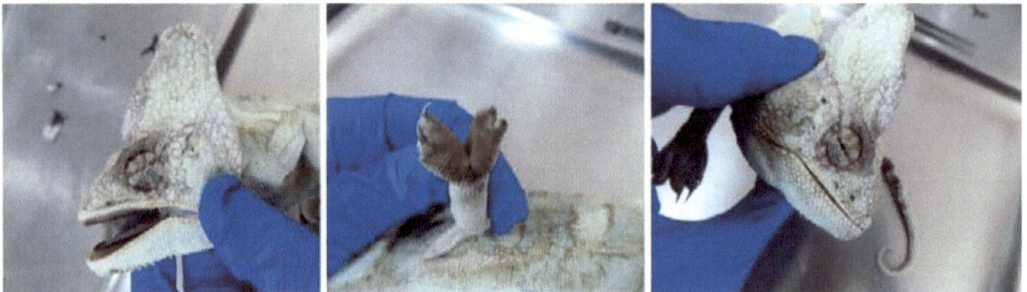

Fig. (13). External exam in a *Chamaeleo calyptratus*.

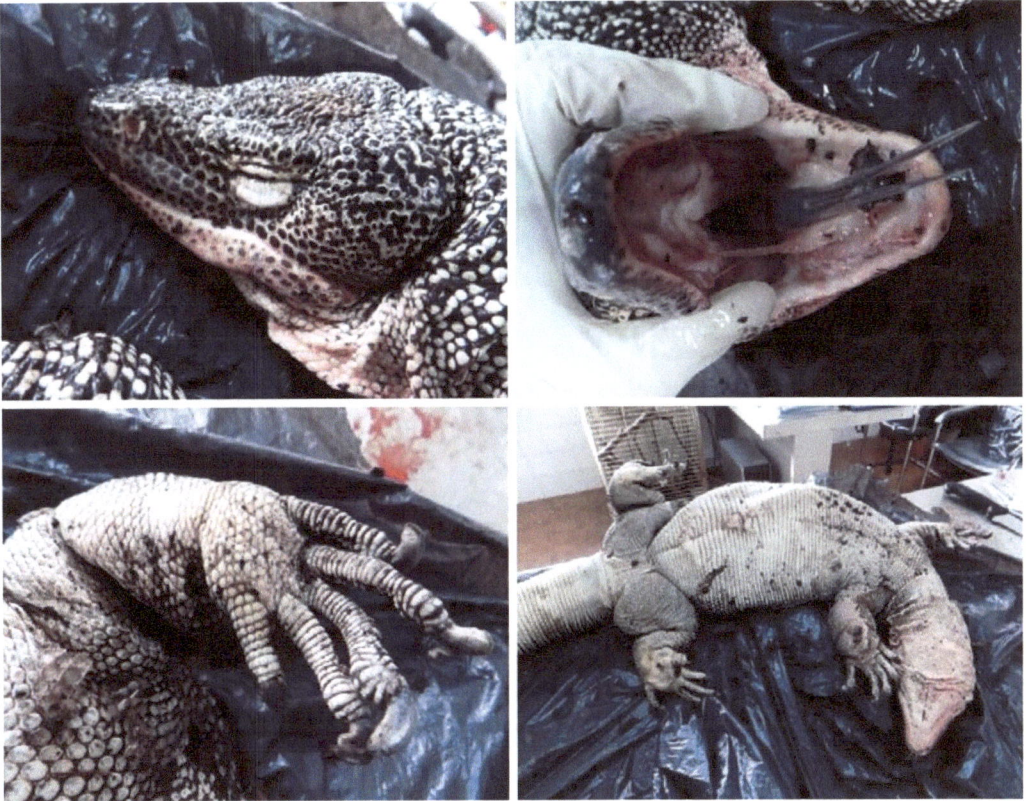

Fig. (14). External exam in a *Varanus indicus.*

The morphology of the tongue varies with the species. Some, such as Monitors and Tengus have bifurcated tongues, chameleons have very long and sticky tongues while the rest have flat and movable tongues (Jacobson, 2007). In the case of Iguana, the end of the tongue presents a darker colouration than the remaining parenchyma, which should not be confused with an injury. The other natural orifices, such as eyes, nostrils, ears, and cloaca should be observed (Figs. **12 – 14**) (Divers and Mader, 2005).

The body condition can be determined by the protuberances of the spinal processes of the vertebrae and the development or atrophy of the muscular masses (Figs. **15 & 16**) (Jacobson, 2007). Also, the fat present at the base of the tail can be a good indicator of nutritional status. Depression of the eyes can also indicate nutritional deficiencies or dehydration (Woodford, Keet and Begins, 2000; Girling and Raiti, 2004).

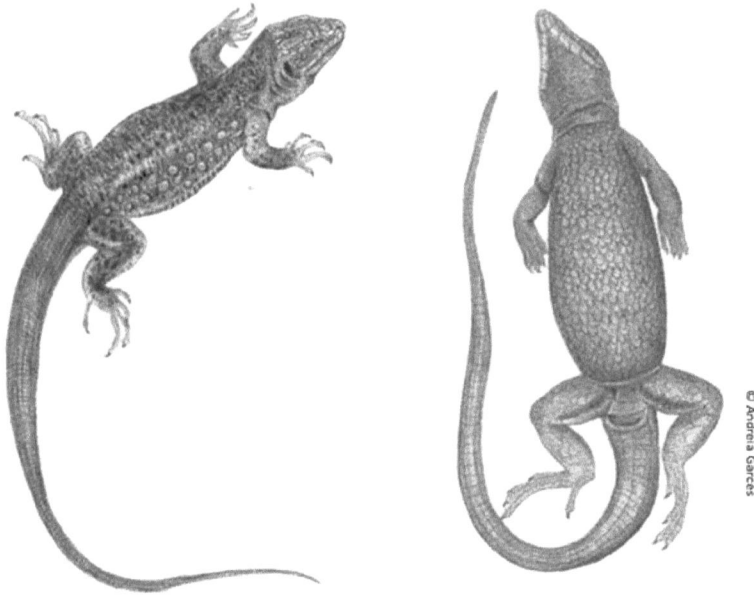

Fig. (15). Representation of the external examination of a lizard.

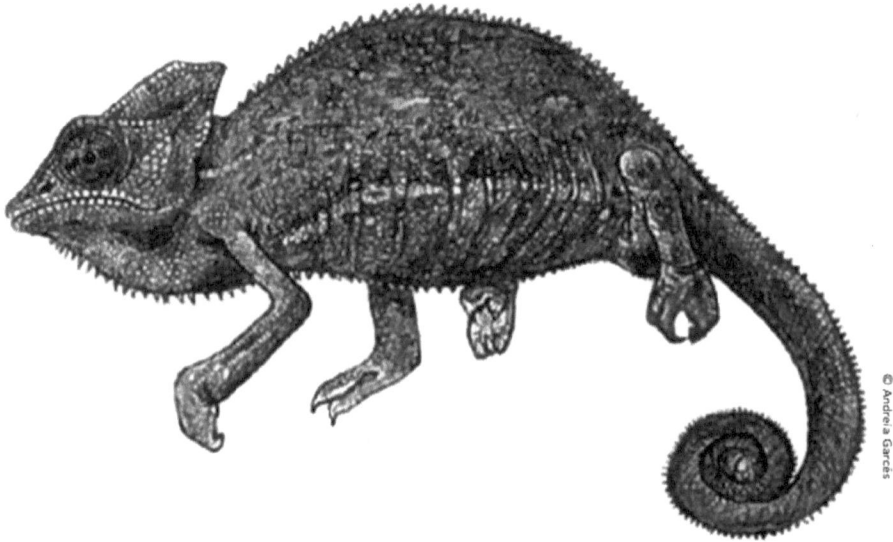

Fig. (16). Representation of the external examination of a chameleon.

Internal Exam

The general conformation of these animals varies with the species. Most present a dorsoventral conformation while others have lateral confirmation, as is the case with chameleons. The method of access to the coelomic cavity will vary with the animal conformation (Terio, Macloose and Leger, 2018).

Animals with dorsoventral conformations should be placed in a ventrodorsal position. An incision is made with scissors or a scalpel, from the cloaca to the intermandibular space following the trajectory of the midline (Fig. **17**). The incision should be performed a little further to the left or right of the midline once, seeing as underneath it there is a large vein, called the ventral abdominal vein (Jacobson, 2007).

The skin is then laterally removed to expose the subcutaneous tissue and muscles. In small animals, the coelomic cavity can be accessed through a section in the muscles, through the midline, from the cloaca to the sternum. The muscles should be removed, and an incision is made in the cartilaginous portions of the ribs to remove the sternum (Munson, 2000; Girling and Raiti, 2004; Divers and Mader, 2005).

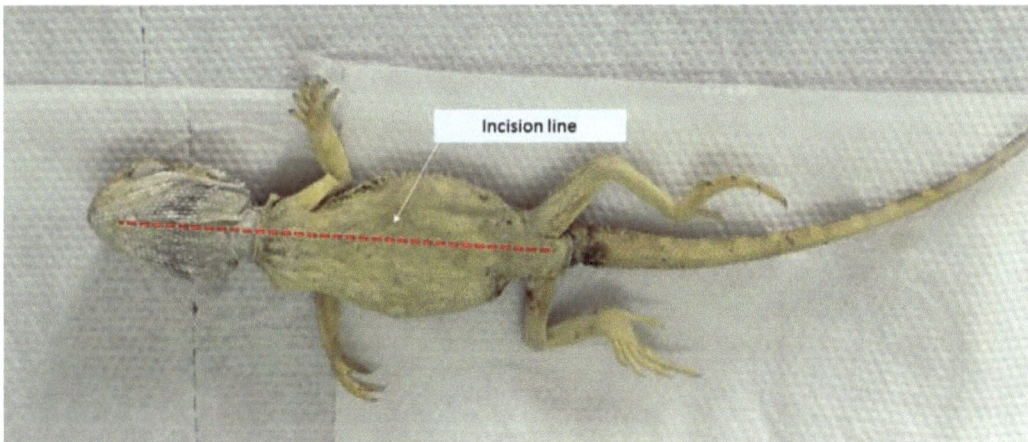

Fig. (17). Representation of the incision line in a lizard.

Animals with lateral conformation should be placed in a lateral position. The limbs that lie on the dorsal side, should then be removed through a cut in their medial face, sectioning muscles and tendons. An incision is made in the midline from the cloaca to the anterior limb, and then the incision is continued dorsally until it returns to the cloaca (Fig. **18**). The muscle and ribs are cut, in order to remove everything in a block and to accede to the coelomic cavity. Cranially, the skin in the lower jaw area should be folded to expose the oral cavity and remaining viscera (Jacobson, 2007).

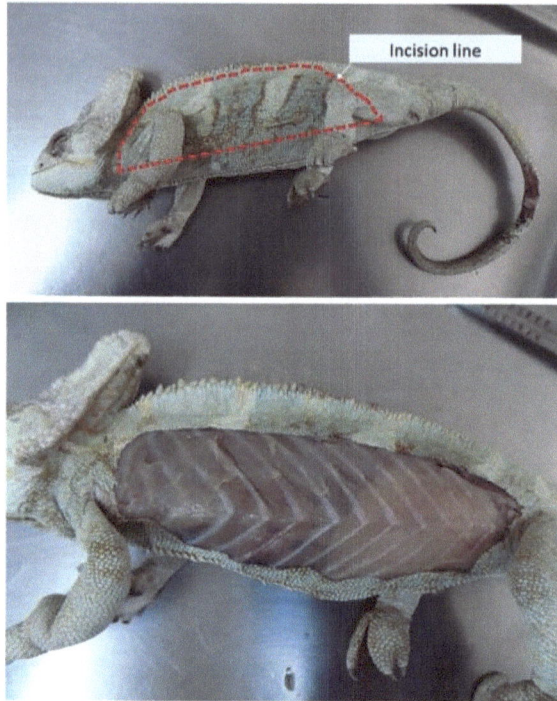

Fig. (18). Line of the incision line and exposition of the ribs on a chameleon (*Chamaeleo calyptratus).*

In the neck area, after removal of the skin, bilaterally is possible to observe the thymus, thyroid, and parathyroid glands. The thymus is located ventral and medial to the internal carotid and jugular vein (Munson, 2000; King *et al.*, 2014). The thyroid can be unique, bilobate depending on the species, as well as the parathyroid glands. In the *Iguana iguana*, the caudal pair is located at the beginning of the internal and external carotid arteries and the anterior glands in the medial area of the mandibular branch (Mader, 2006).

After the opening of the coelomic cavity, all the organs should be observed *in situ*. The presence of free fluid, fibrous or purulent material in the coelomic cavity should be recorded. There is intracellomic fat in these animals that must be removed before the organs can be observed. It also provides information about the animal's nutritional status. In chameleons and some species of lizards, a melanic pigmentation can be observed in the mesenteric and inner surfaces of the celomic wall and should not be confused with a pathological alteration (Jacobson, 2007). The organs of the gastrointestinal tract can be removed as a single block by making an incision on the cranial oesophagus and another in the most distal possible near the cloaca. Afterwards observe one by one he organs at the necropsy table (Figs. **19-20**) (Chai, Girling and Raiti, 2004).

The examination is begun by the cardiovascular system *in situ*, observing its shape, size and colouration *in situ* (Jacobson, 2007). The opening is similar to what is described for animals of the Testudinata Order, including the presence of a *gubernaculum cordis* (Girling and Raiti, 2004).

The respiratory system is then observed. The examination of the pharynx and the trachea begins. The trachea has incomplete cartilaginous rings. An incision is made along the trachea with the aid of scissors.

The lung in reptiles resembles a sac that, when cut, is identical to the hexagons of a beehive. The thicker-wall cranial portion is responsible for respiratory function and, as such, can provide us with more information. Concerning the anatomy of the lung, it is possible to find three types: the most primitive with a single chamber (present in the families Iguanidae and Agamidaee) and the most evolved with multiple chambers (we can observe in Varanos and Helodermatides) (Jacobson, 2007). In chameleons, the lungs are of the paucicameral type, where we can see a tentacular diverticulum projecting from the lungs, similar to an air sac (Fig. **20**). They should be palpated carefully and sectioned transversely to detect any parasites or exudates. Foci of pigmentation may also be normal (Munson, 2000; Meredith and Redro, 2002; Divers and Mader, 2005).

Next, the gastrointestinal tract and the attached glands should be observed. In normal situations, the liver presents a homogeneous colouration of a dark brown tone (Jacobson, 2007). Its examination is similar to what has been described previously for the snakes. The gallbladder may or may not be present caudally to the liver. This should be sectioned to observe the contents and mucosa (Divers and Mader, 2005).

The oesophagus, stomach, small intestine and large intestine should be visually examined, palpated, and several segments collected for further histopathological examination. The gut serosa, in some species of lizards, may be pigmented. Herbivorous species, such as the Iguana, have a well-developed cecum and sacculations (Jacobson, 2007). Using a pair of scissors, they should be cross-sectioned and their internal mucosa should be examined, as well as the contents, which should be collected whenever poisoning is suspected. The pancreas is located near the duodenal loop (Munson, 2000; Viner and Kagan, 2018). This should be observed as fresh as possible as it degrades quickly.

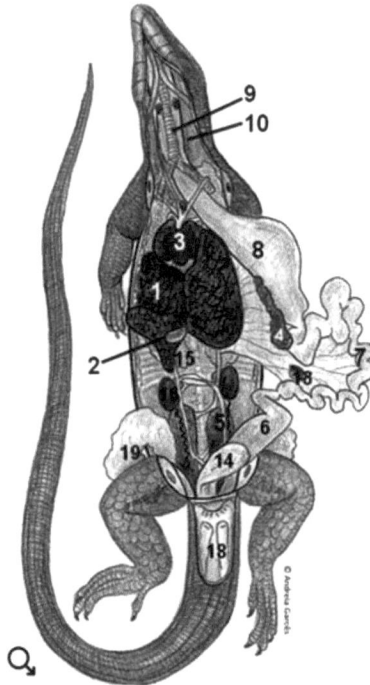

Fig. (19). Schematic representation of the organs *in situ* in a male lizard. (**1**-liver, **2**-gallbladder, **3**-heart, **4** pancreas, **5**-kidney, **6**-large intestine, **7**-small intestine, **8**-stomach, **9**-trachea, **10**-esophagus, **11**-thyroid, **12**-parathyroid **13** - spleen, **14** - bladder, **15** - lung, **16** - testicles, **17** - adrenal gland, **20** - ovary, **21** - oviduct).

The spleen is ovoid and dark red and is associated with the pancreas and stomach (Jacobson, 2007). In varans, we can normally observe portions of the endocrine pancreas in the spleen (Divers and Mader, 2005).

The kidneys, gonads and adrenal glands are found in the caudodorsal region of the coelomic cavity. In some species it is located in a retrocelomic position, in the pelvic canal, being necessary to cut the pelvic bone to access them (Fig. **20**). The kidneys have an ovoid shape and should also be evaluated (Jacobson, 2007). In sexually active males, they exhibit sexual segment formations in the kidneys, which are represented as areas of prominent white colouration, which should not be confused with pathological changes (Jacobson, 2007). They may or may not have a urinary bladder present. The adrenal glands are in a cranial (females) or medial (males) position to the kidneys. They are elongated in pairs and their colouration may be yellowish or white (Mader, 2006).

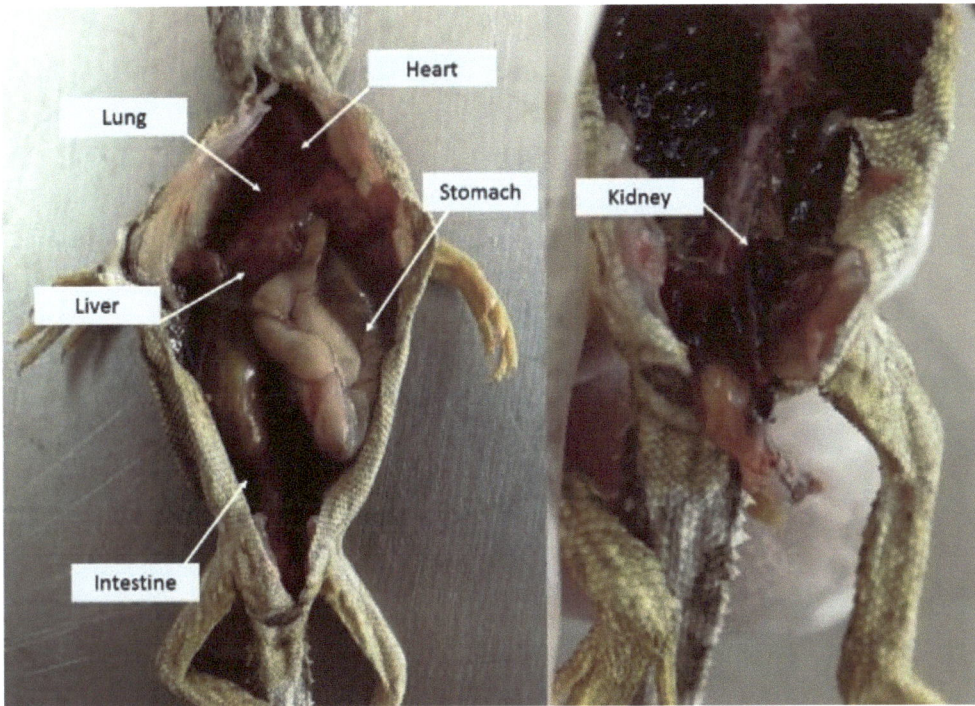

Fig. (20). Internal exam in a *Pogona vitticeps.*

The gonads have an intracellular location near the kidneys. Females may be oviparous or viviparous (Jacobson, 2007). The ovaries and oviducts are located in a cranial position relative to the kidneys, and later into the uterus and vagina, and finally into the cloaca through which the embryos are expelled (Girling and Raiti, 2004). In males, the testes have an oval shape and are asymmetrical. In the ventral region of the tail zone, they have two hemipenes. These organs should be removed and observed for any possible changes (Figs. **21-23**) (Divers and Mader, 2005).

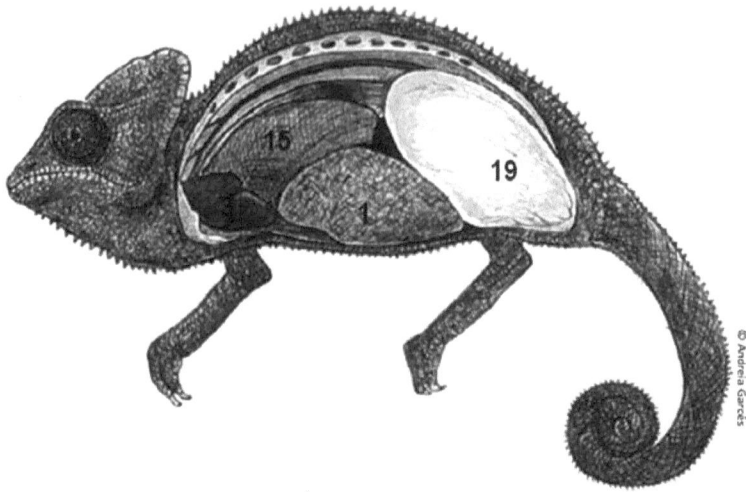

Fig. (21). *In situ* organs in a chameleon after removal of skin and muscles (1-liver, 15-lung, 19-fat).

The tail should be cut at its base and removed so that the fat deposits between the dermis and the deep muscles are visible (Fig. **21**) (Jacobson, 2007). Some species can amputate the tail in situations of stress or self-defence, so it is possible sometimes to observe its regeneration in several different stages. The joints of the limbs should be also sectioned (Divers and Mader, 2005).

To observe the brain the animal should be decapitated at the level of the atlas. Then make an incision on the skin and move it away to expose the skull (Jacobson, 2007). Then, depending on the size of the animal, the skull can be opened with a pair of scissors or a bone saw. Likewise, cross-sections can be performed on the spine to observe the bone marrow and spinal canal and take samples if necessary (Divers and Mader, 2005).

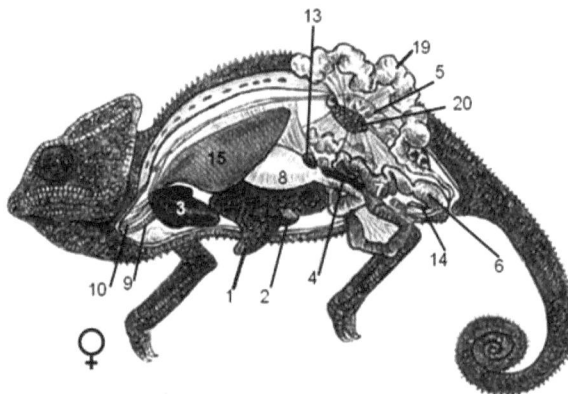

Fig. 22 cont.....

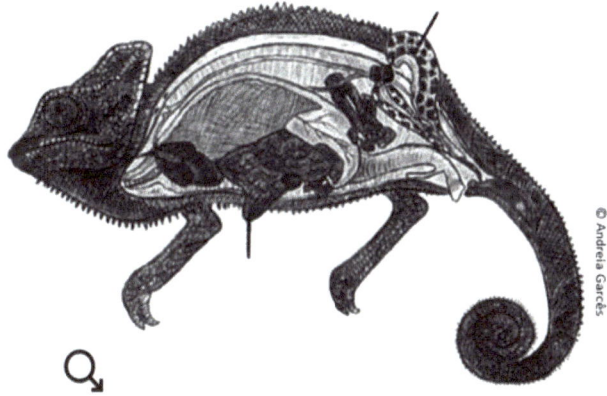

Fig. (22). Schematic representation of the *in situ* organs in a male and female chameleon. (**1**-liver, **2**-gallbladder, **3**-heart, **4**-pancreas, **5** kidney, **6**-large intestine, **7**-small intestine, **8**-stomach, **9**-trachea, **10**-esophagus, **11**-thyroid, **12**-parathyroid **13** - spleen, **14** - bladder, **15** - lung, **16** - testicles, **17** - adrenal gland, **19** - fat, **20** ovaries, **21** - oviduct).

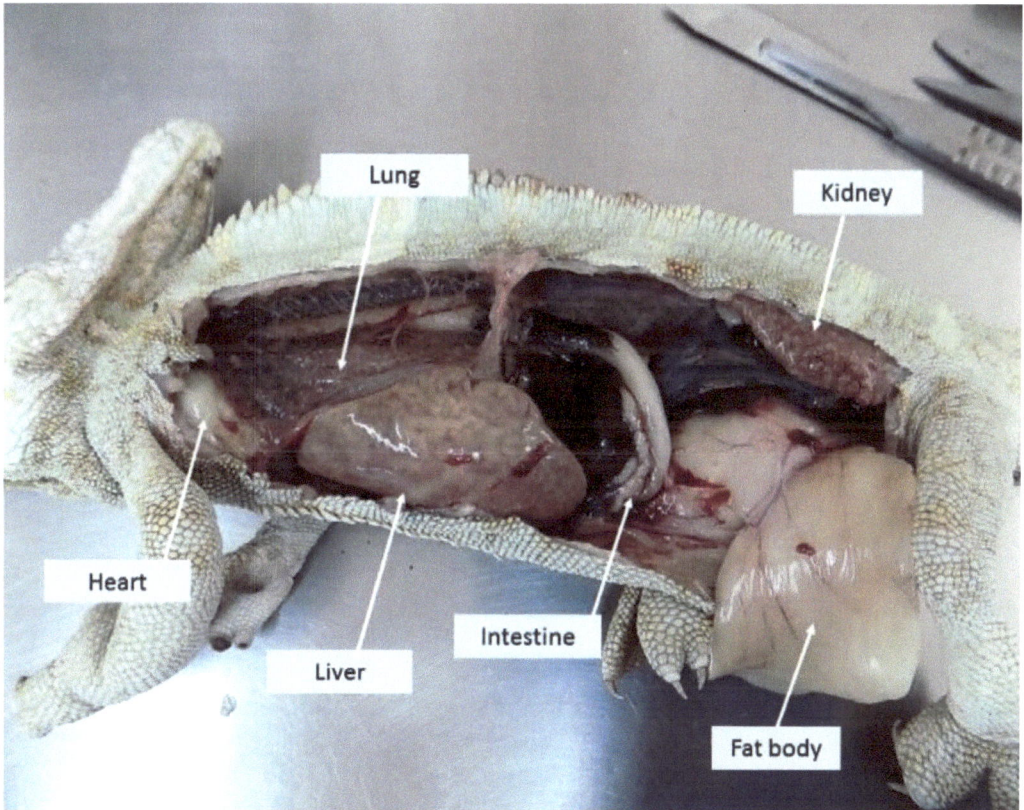

Fig. 23 cont.....

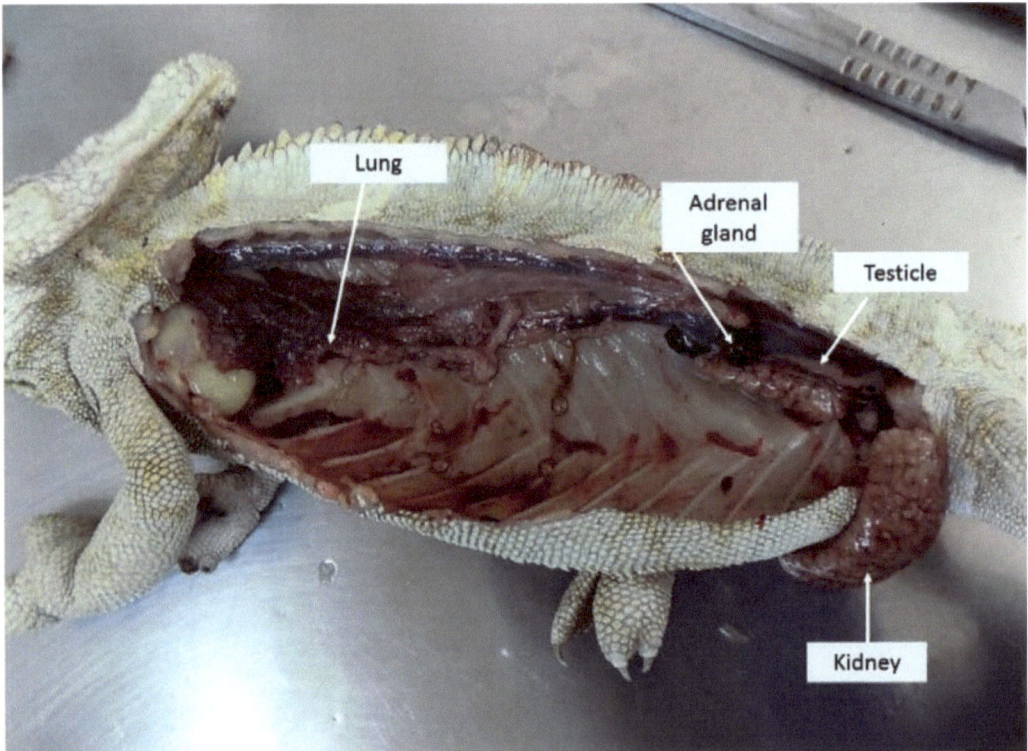

Fig. (23). Internal exam in a *Chamaeleo calyptratus.*

ORDER CROCODYLIA

General Considerations Before the Necropsy

The size of these individuals is very variable, from a few centimetres with newborns up to several meters in adults (Fig. **24**). There are 23 species of crocodiles classified in three families: Alligatoridae, Crocodylidae and Gavialidae (Fig. **25**) (Divers and Mader, 2005; O´Malley, 2005).

In order to preserve the microbial flora of these animals for microbiological examination, attention should be paid to the fact that refrigeration does not prevent the multiplication of psychotropic bacteria, which often mask the presence of other pathogenic bacteria. So, the *post-mortem* exam should be done as fast as possible after their death (Huchzermeyer, 2009).

External Exam

The external exam should begin by identifying the species and subspecies if possible. The carcass should then be weighed and measured using biometric measures described in the existing bibliography (Jacobson, 2007).

Fig. (24). *Caiman crocodilus.*

Crocodile skin has an armed integument, especially on the dorsal portion of the body. In the area of the head and back these shields contain bony plates, denominated osteodermas. They do not possess sweat glands in the skin. Crocodiles may have some gular glands in the ventral jaw and some in the lips and cloaca (Jacobson, 2007). In *Crocodylus niloticus* and *Alligator sinensis,* we can observe rudimentary glands on the dorsum between the second and fifth rows of scales (Munson, 2000; Divers and Mader, 2005; Huchzermeyer, 2009).

The skin should be examined for observing the presence of traumatic lesions, scars, changes in pigmentation, scaling asymmetries, ulcers, Pox Virus lesions, abrasions (especially in the cheek area and in the plantar area of the limbs) ectoparasites or other lesions (Jacobson, 2007). At this stage pieces of skin can be removed for histopathology examinations (Meredith and Redro, 2002; Girling and Raiti, 2004; Divers and Mader, 2005).

The carcass should be palpate to observe the presence of fractures, dislocations, masses, oedema or other abnormalities. In the limbs, it is also possible to observe the presence of amputations, arthritis, lack of digits or pododermatitis (Jacobson, 2007).

In the eyes, observe the eyelids, the nictitating membrane (well developed in these animals to protect them when they plunge) and the cornea. The teeth should also be observed. Count them and compare them with the dental formulas. (Jacobson, 2007). The teeth in this order are replaced throughout the life of the animal, with smaller intervals when they are young. The most common lesions possible to observe in the mouth are broken teeth, ulcers, fistulas, abscesses, traumatic lesions, chemical burns, inflammation, foreign bodies, among others (Mader, 2006; Huchzermeyer, 2009).

The tympanic membrane is located in a depression behind the eyes and is protected by a leaflet of the integument that closes whenever they dive (Jacobson, 2007). The other orifices such as nostrils and cloaca should also be observed (Huchzermeyer, 2009).

Família *Crocodylidae*

Família Alligatoridae

Fig. 25 cont.....

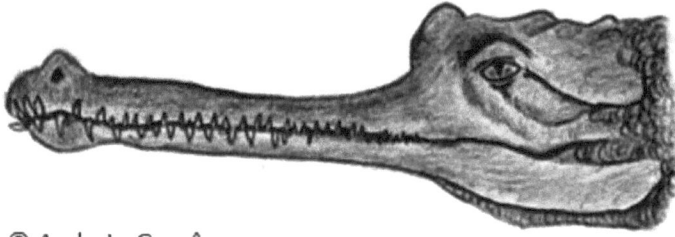

Família Gavialidae

© Andreia Garcês

Fig. (25). Differences in the three families in the Order Crocodylia.

Body condition may be somewhat more difficult to determine in these animals due to the presence of bone plaques along the dorsal portion of their body (Jacobson, 2007). In these animals, the fat present at the base of the tail can be a good indicator of the nutritional status. Depression of the eyes can indicate nutritional deficiencies or dehydration (Munson, 2000; Terio, Macloose and Leger, 2018).

Internal Exam

To begin the procedure, the carcass should be placed in a ventrodorsal position, with the abdominal area facing upwards. An incision is then made between the anterior limbs, cranially to the row of large plaques between them. The incision should continue, caudally, on both sides underneath the front limbs toward the hind limbs. Care must be taken not to damage any abdominal organs, especially the stomach, which is adhered to the left abdominal wall. Then these lateral incisions should be continued cranially, towards the junction of the maxilla. On the pectoral muscle, the coracoid and the cartilaginous portions of the ribs should be removed together with the skin in a single block. The coracoid is quite dense, and in larger animals, it is easier to cut thought its cartilaginous junction in the midline (Jacobson, 2007). Another transverse incision should be performed in the caudal portion, where the protuberance of the pubic bone is located. In adult animals, the incision is performed cranially to this structure and in very young animals and newborns this bone can be easily removed, and the incision can be performed directly on the bone. Raise the skin, with all the adhesions carefully dissected, to expose the coelomic cavity (Fig. **26**) (Girling and Raiti, 2004; Divers and Mader, 2005; Huchzermeyer, 2009).

The incision that has been extended cranially towards the maxillary junction, should be deepened, medially, from the back of the jaws to the tip of the tongue. This flap raised with the tongue, exposing the soft palate, the internal nostrils, the ventral opening of the eustachian, the pharynx tubes and the dorsal leaflet of the gular valve. The region of the glottis and gular valve should be closely observed since they are sites where lesions can be observed when present (Jacobson, 2007). In the neck area, after removing the skin, it is possible to observe bilaterally (with some fat when the animal is in a good body condition) the thymus, thyroid and parathyroid gland. Depending on the species, it may be possible to observe one or two pairs of parathyroid glands located along the common carotid artery. Some species of crocodiles have a single lobulated thyroid while other species have two separate lobules on each side of the trachea, with a reddish-brown colouration. An exception is the Nile Crocodile (Jacobson, 2007), which has the thyroid on the lateral side of the bronchi and medially to the carotid artery in conjunction with the parathyroid. The thymus in crocodiles never recedes with age and is located throughout the neck to the base of the heart, but in emaciated animals, it may not be very easy to perceive (Munson, 2000; Girling and Raiti, 2004; Divers and Mader, 2005).

Fig. (26). External habit of a crocodile.

After opening the coelomic cavity, all the organs should be observed *in situ*. The presence of free fluid or fibrous or purulent material in the coelomic cavity should be recorded (Jacobson, 2007).

In this stage observe the shape, size and colouration of the hearth *in situ*. The pericardial sac should be opened to assess if there is the presence of fluid which, in reduced amounts, should be considered physiological. Its colour, quantity, or presence of other substances such as blood or fibrin should be recorded (Huchzermeyer, 2009).

When opening the pericardium observe the *gubernaculum cordis fibrosus*, which connects the apex of the heart to the sac of the pericardium, (normal structure in crocodiles). The heart is removed, trying to leave as much of the great vessels intact as possible. They are the only group of reptiles whose heart contains 4 chambers (two atria and two ventricles). They do not possess any fat in the coronary sulcus and their interventricular septum is complete. Besides, they have three cardiac adaptations to submerge: a right and left artery, connective tissue from the lungs to the ventricle that allows a restricted flow of blood during dives and the *panizza foramen* (opening between the right and left aortic arch) (Jacobson, 2007). The heart should be palpated and visually examined for its shape, colour, or presence of any injuries such as trauma, mass or foreign bodies. Then the chambers should be opened to observe the myocardium and cardiac lumen. The large vessels, the left and right arteries should be inspected, detect changes such as arteriosclerosis or dystrophic mineralization (Divers and Mader, 2005; Huchzermeyer, 2009).

The respiratory system should then be observed. This system as already been partially observed with the inspection of the nostrils (which have muscles that enclose them during submersion) and the gull valve. The trachea in these animals has complete cartilaginous rings (Jacobson, 2007). An incision should be made along the tracheal tube with the aid of a pair of scissors, in order to observe the presence of changes in the mucosa and presence of abnormal contents (liquid, food, foreign bodies, parasites) in the tracheal lumen. In adults, the trachea bifurcates into two very extensive extrapulmonary bronchi that should be closely examined to detect the presence of lesions and exudates (Jacobson, 2007).

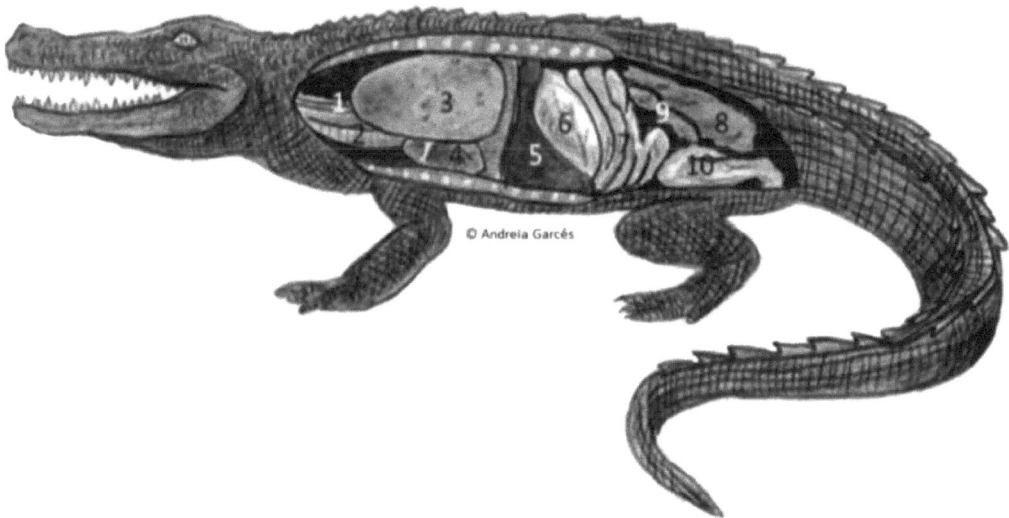

Fig. (27). Schematic representation of in situ organs in male crocodile (**1**-esophagus, **2**-trachea, **3**-lung, **4**-heart, **5**-liver, **6**-stomach, **7**-intestine, **8**-kidney, **9**-gonads).

The lungs in this species are well developed and the vessel wall is thicker when compared to mammals. Crocodiles possess a post-pulmonary membrane that separates the lungs from the liver and a post-hepatic membrane that is attached to a strip of muscle that penetrates onto the pubis (Jacobson, 2007). Together, these two membranes act as a diaphragm. The two lungs should be removed from the celomic cavity by cutting their connections to the celomic wall and the pre-hepatic transverse membrane (Jacobson, 2007). The lungs should be palpated carefully to assess changes inconsistency or the presence of other lesions. Transverse sections in the parenchyma should be performed to detect the presence of parasites or abnormal exudates. Cut the remaining membranes for easier access to the remaining viscera (Fig. **27**) (Divers and Mader, 2005; Huchzermeyer, 2009; Terio, Macloose and Leger, 2018).

The liver and stomach are located on the left side of the celomic cavity and the fat bodies and some intestinal loops on the right side. These fat bodies are also useful to indicate more accurately the nutritional status of the animal. When the animal was chronically sick or aporetic this structure is atrophied (Jacobson, 2007). After removing the fat bodies is possible to observe the spleen, which is located dorsally to the mesenteric. The organs of the digestive system should be removed in a single block by making an incision on the cranial portion of the oesophagus and another near the cloaca, as distal as possible (Huchzermeyer, 2009).

The oesophagus, stomach, small intestine and large intestine should be examined visually and palpated to observe the presence of foreign bodies or masses. Then, using a pair of scissors, they should be opened to observe their lumen and detect if there is the presence of lesions such as gastric ulcers, traumas, masses, or other lesions (Jacobson, 2007). The mucosa of the small intestine in these animals is very thick and muscular, and this should not be confused with some pathology. Gastrointestinal contents may be collected whenever poisoning is suspected (Divers and Mader, 2005; Huchzermeyer, 2009).

The pancreas surrounds the spleen caudally, while cranially is located between the loops of the small intestine. It should be observed as fresh as possible as it degrades quickly (Huchzermeyer, 2009).

The spleen is an ovoid organ with a dark red colouration. Its parenchyma is surrounded by a thick capsule, which usually has the elasticity to grow with the organ. If this does not occur, observe the parenchyma and capsule, may be indicative of an infection or septicaemia (Jacobson, 2007).

The liver has a dark brown colouration and is divided into two lobes. The gallbladder is located between the two lobes. These should be visually examined and palpated to identify abnormal structures (Jacobson, 2007). In the liver, the parenchyma sectioned to observe if there is the presence of parasites, zones of necrosis, abscesses or alteration of the walls of the bile canaliculi and blood vessels. The gallbladder should be open to observe the lumen, contents and detect obstructions or calculations (Divers and Mader, 2005; Huchzermeyer, 2009; Garcês and Pires, 2017).

The kidneys, gonads and adrenal glands are located in the caudodorsal region of the coelomic cavity. The kidneys have an ovoid and should be examined to detect the presence of urates, chronic kidney disease, urolithiasis, bacterial infections, parasites or neoplasms (Jacobson, 2007). Crocodiles do not have a urinary bladder, so the ureters drain directly into the urodeo that should be examined to detect the presence of stones or masses. The adrenal glands are elongated structures located in the cranial region of the kidneys. Its colouration may be yellowish or white (Jacobson, 2007; Huchzermeyer, 2009).

Females are oviparous. The ovaries are located cranial to the kidneys. The ovaries have an elongated shape and, in very young animals, are difficult to differentiate from the male gonads (Jacobson, 2007). In adults, the follicular structure becomes quite differentiated, and the follicles all differentiate at the same time. The oviducts become thicker with maturity and sexual activity, entering the uterus (glandular portion), followed by the vagina. In the vagina, the eggs are stored and then expelled through the cloaca (Jacobson, 2007).In addition, ventrally to the cloaca, they have a small clitoris. In males, the testes are located in a cranial position to the kidneys, supported by the meschoria. In the ventral region of the tail, they have a penis. Normally it is not possible to differentiate the sex, except in the genus Gharial (Jacobson, 2007), that has a bulbous structure in the rostral zone of the upper jaw that develops when they are approximately 13 years old. These organs should be removed and observed separately in the necropsy table to observe the presence of changes such as neoplasms, abscesses or dystocia (Jacobson, 2007; Huchzermeyer, 2009).

The tail should be cut at its base and removed to observe the fat deposits between the dermis and the deep muscles. The joints of the limbs should be sectioned to observe the presence of arthritis, gout, and other lesions.

To observe the brain, in large animals, make an incision at the level of the atlas, and then with a bone saw cut the scales and skull bone directly (Jacobson, 2007). In small animals (less than 1,5m), the brain is accessed from the ventral surface of the skull. To do this, the skull is placed in a ventrodorsal position, and with a knife and a hammer, an incision is made in the midline of the palate (Jacobson, 2007). Consequently, the operator also has access to the nasal cavities and both cerebral hemispheres. Likewise, transverse cuts can be made in the spine to observe the bone marrow and spinal canal and withdraw samples if necessary (Munson, 2000; Mader, 2006; Huchzermeyer, 2009).

ORDER TESTUDINATA

General Considerations Before the Necropsy

There are about 322 species of chelonians - turtles, turtles and sea turtles. They can live in terrestrial, aquatic and marine environments. Its size varies from species like *Homopus signatus* with the only 100g to animals of the species *Dermochelys coriacas* that can reach up to 800kg (Mader, 2006).

This order is characterized by the animals having a carapace. They have the evolutionary capacity to introduce the head inside the carapace to protect themselves from predators. Some species do not have this characteristic like the sea turtles and the genus Chelydridae (Jacobson, 2007).

The chelonians can be divided into two orders depending on how they flex their cervical vertebrae (Mader, 2006): Pleurodira (aquatic and semiaquatic, which fold the neck in am horizontal "S" with the head sideways) and Cryptodira (most species, neck folds in vertical "S", with the head completely inside the carapace) (Jacobson, 2007).

External Exam

The external examination begins with the identification of the species. Body condition is difficult to evaluate because of its anatomical characteristics. One way is to weight the animal and compare it with the normal parameters described to that species. Another way is to use a body condition index, but it only exists for some species of chelonians (Munson, 2000; Survey, 2000; Flint *et al.*, 2009; Farris *et al.*, 2014).

The skin should be observed to detect the presence of ulceration, erythema, exudates, oedema, traumatic lesions, pigmentation changes, scars, abrasions or masses. At this stage, we can remove fragments of skin for histopathological examination(Mader, 2006).

Regarding the carapace and the plastron, it should be visually observed to detect the presence of any lesions such as ulcers, haemorrhages, fractures, changes in the conformation of the carapace (triangular shape for example) (Fig. **28**) (Mader, 2006), erosion of the shields or changes in colouration. The carapace should then be palpated to detect the presence of possible fractures, changes in growth and changes in consistency (*e.g.* hypovitaminosis problems) (Jacobson, 2007).

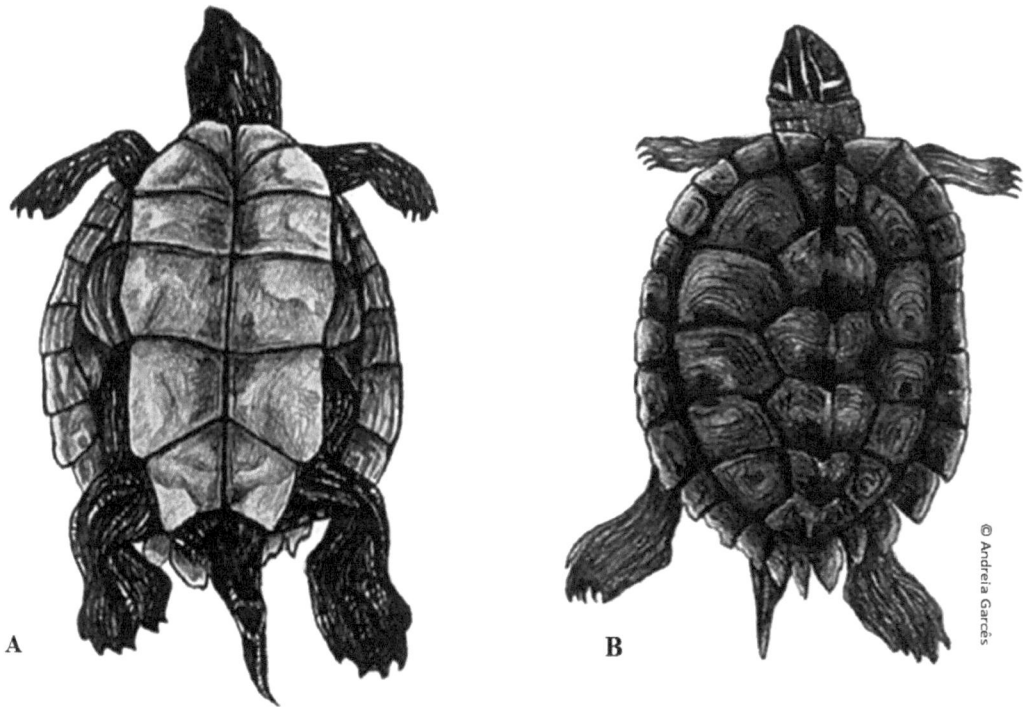

Fig. (28). External habit of a chelonian, in ventrodorsal position (**A**) and dorsoventral (**B**).

In the limbs and head, observe the presence of fractures, traumatic lesions, masses, muscular atrophy, overgrowth or deformation of the nails, oedema, lack of digits, pododermatitis or arthritis, for example (Jacobson, 2007). Observe the presence of ectoparasites that are removed for later identification. In sea turtles, attention should be paid to the amount and distribution of epibiota that is present in the shell, which is an indicator of the animal's health (Mader, 2006).

It is sometimes possible to perform sexing due to some characteristics. In males the plastron has a concavity, the tail is long and thin, and the cloaca is situated beyond the margin of the carapace. In the aquatic turtles, the males have in the thoracic limbs' very long claws (Mader, 2006) (Fig. **29**).

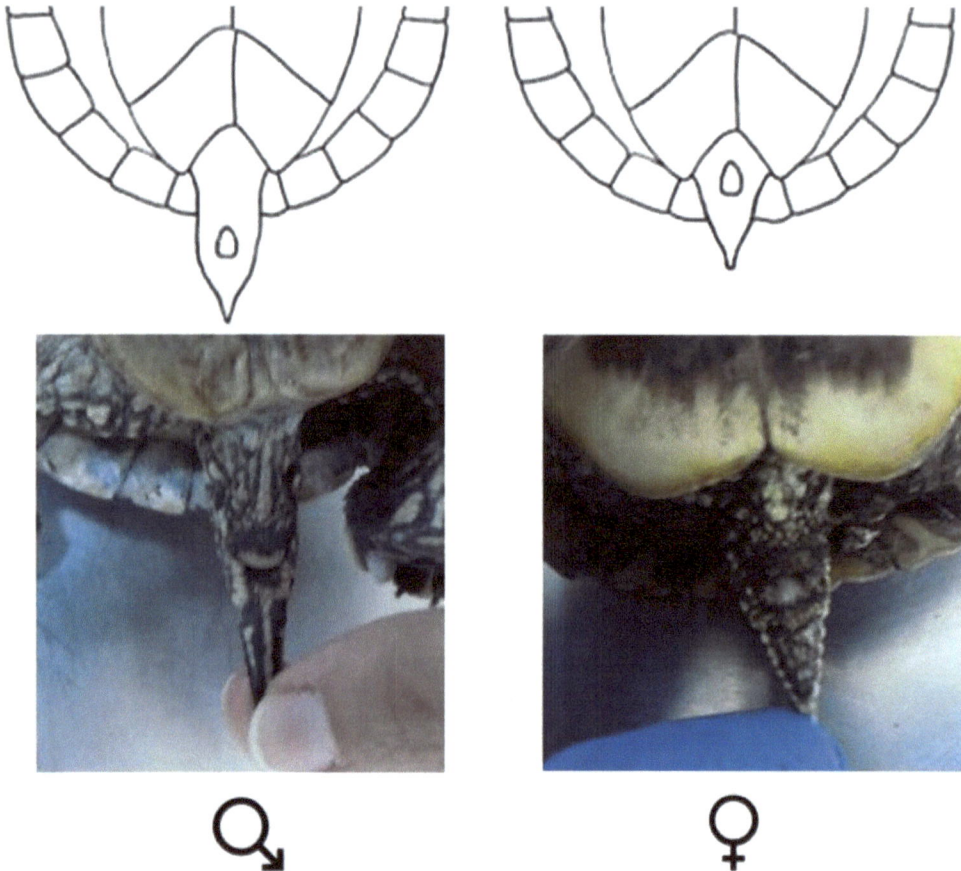

Fig. (29). Sexing of tortoises.

The cloaca, eyes, mouth, ears and nostrils should be observed in order to detect the presence of abnormal secretions, masses or other lesions. The chelonians do not have teeth, but a rather strong and sharp beak, which should be observed to verify the presence of malocclusion of the beak, fractures or other lesions (Munson, 2000; Survey, 2000; Flint *et al.*, 2009; Farris *et al.*, 2014) (Fig. **30**).

The interior of the mouth (oropharynx, glottis and tongue) should be examined to detect the presence of trauma, ulcers, chemical burns, stomatitis or foreign bodies, among others. The pale mucosa is normal in a healthy animal after death. In some diseases, the mucosa may be cyanotic or haemorrhagic (Mader, 2006).

Otitis is very common in these animals. The ears should be examined to observe the presence of oedema or abscesses in the tympanic membrane (Munson, 2000; Survey, 2000; Flint *et al.*, 2009; Farris *et al.*, 2014). In the eyes observe the eyelids, nictitating membrane and cornea. Depression in the eyes can indicate nutritional deficiencies or dehydration (Figs. **31-35**) (Jacobson, 2007).

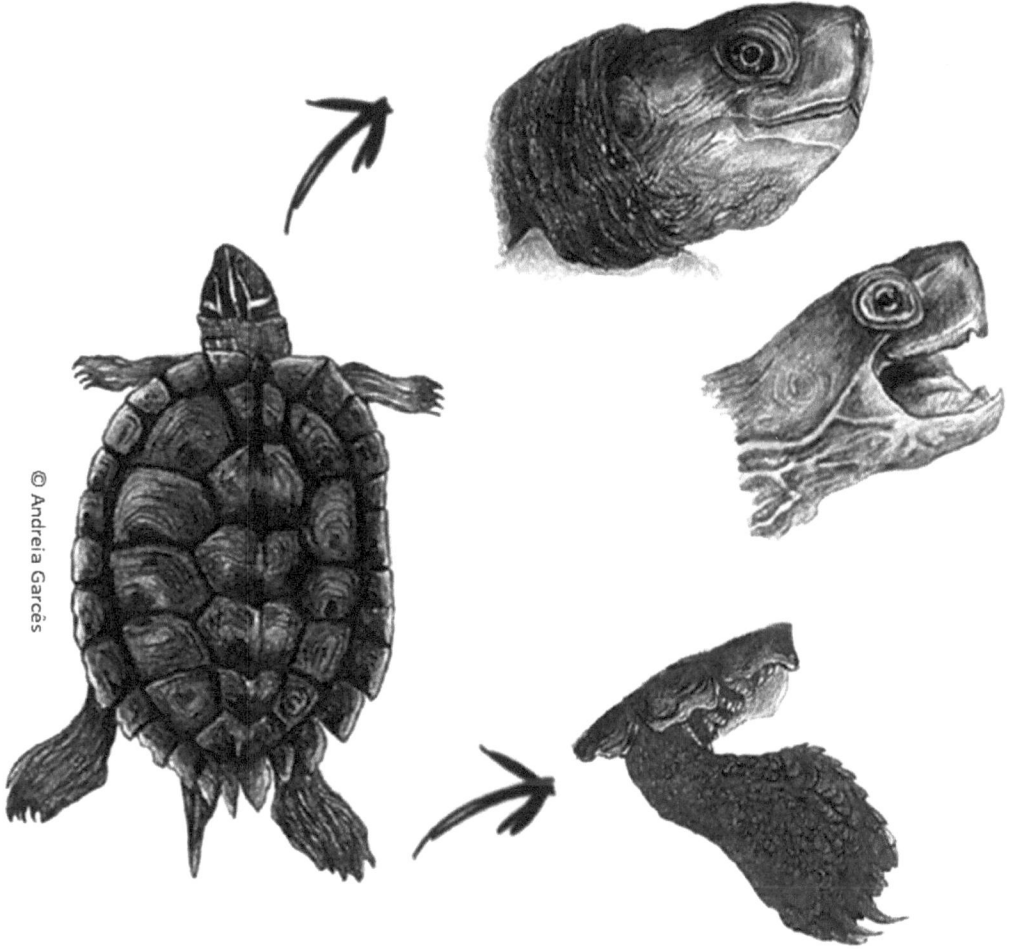

Fig. (30). Detail of external habit in a tortoise.

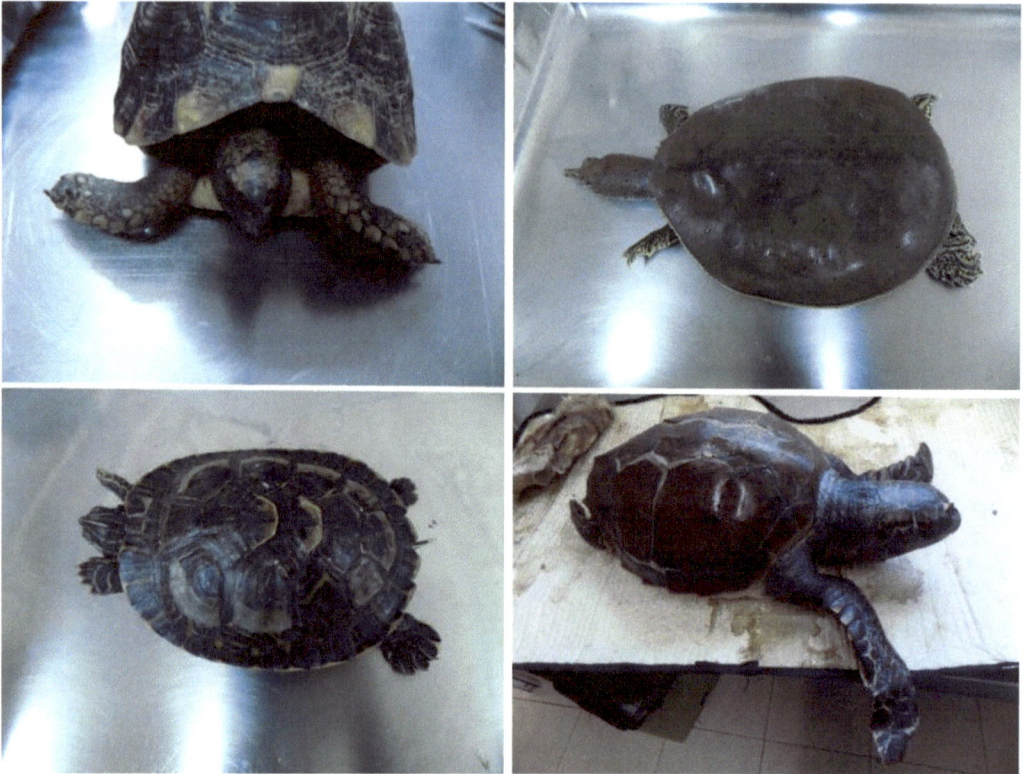

Fig. (31). Representation of different species from the Order Testudinate.

Fig. 32 cont.....

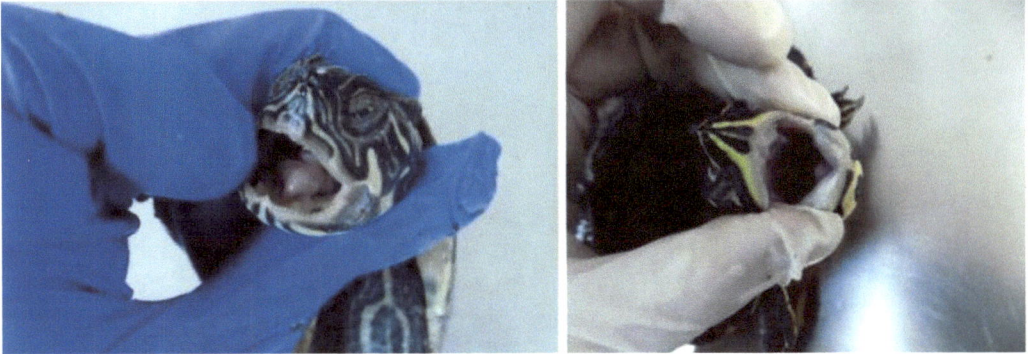

Fig. (32). External exam in a *Pseudemys nelson.*

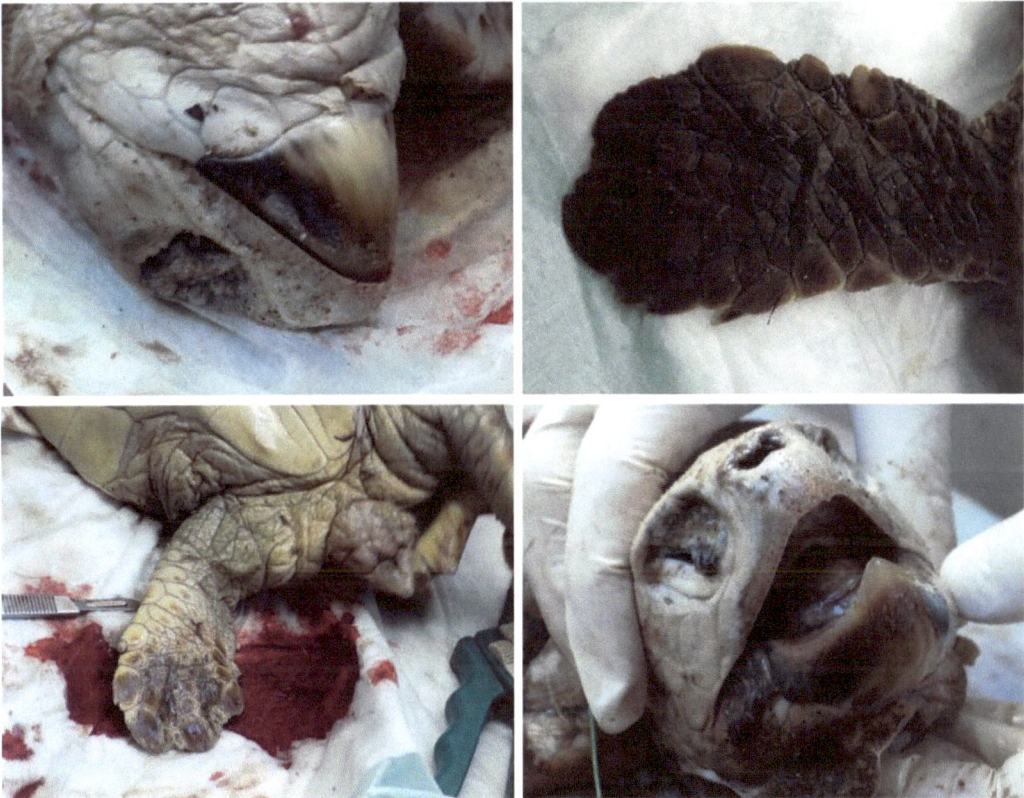

Fig. (33). External exam in a *Caretta caretta.*

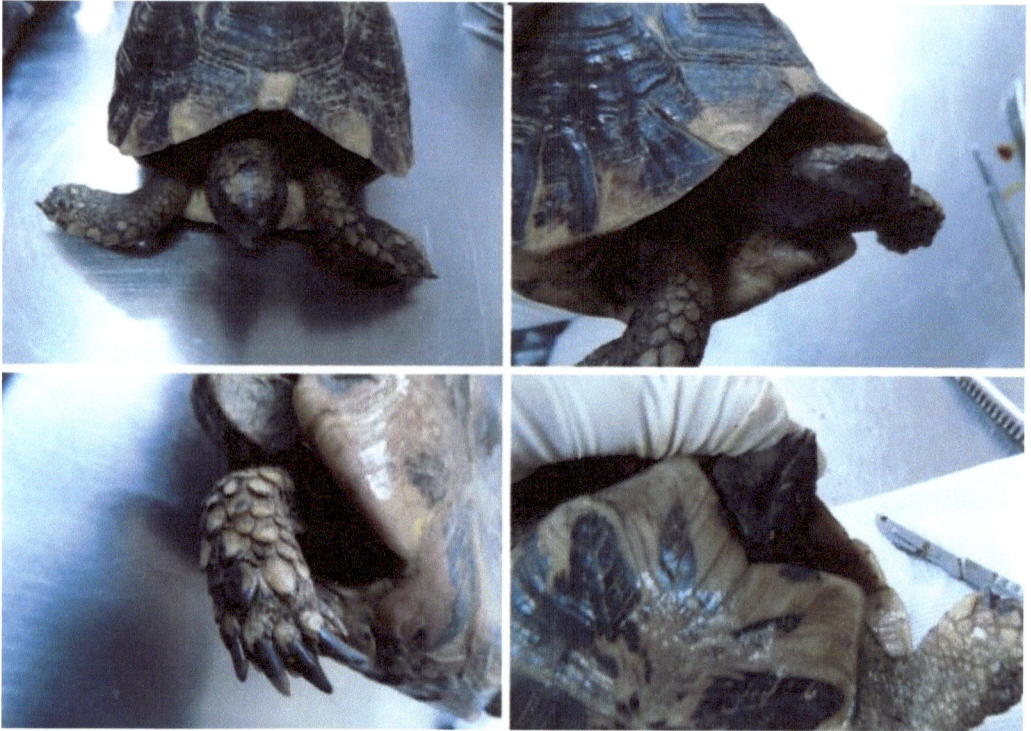

Fig. (34). External exam in a *Testudo graeca.*

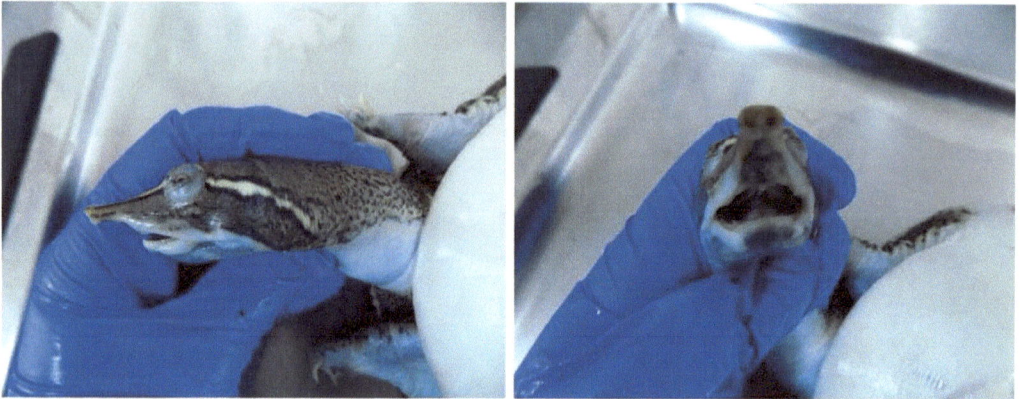

Fig. 35 cont.....

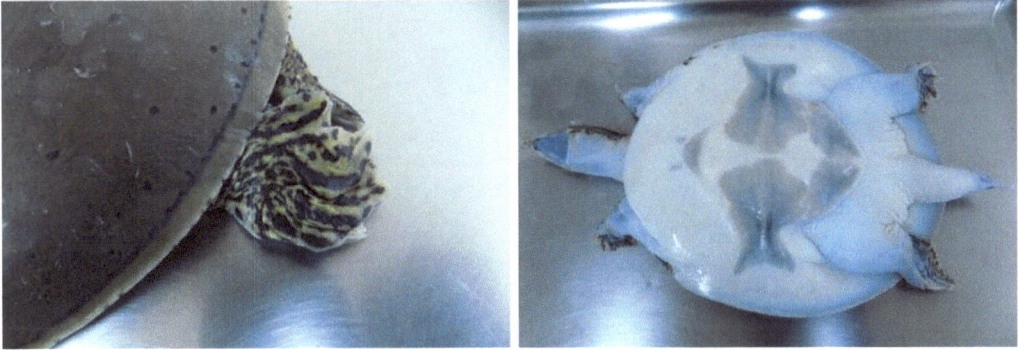

Fig. (35). External exam in a *Pelodiscus sinensis.*

Internal Exam

To initiate necropsy, the animal should be placed in a ventrodorsal position. Due to its anatomical characteristics, specific material is needed to access the coelomic cavity. Depending on the size of the animal, a pair of pliers or a bone saw should be used. The plastron should be removed as a single piece. To do this first is necessary to make an incision on the skin at the edges of the shell with a scalpel, as close to it as possible (Jacobson, 2007). The marginal bridge on both sides between the limbs, connecting the plastron to the carapace, due to its rigidity, must be incised with the bone saw, in a line extending from the front to the posterior limb on both sides(Mader, 2006). Next, elevate the plastron and, with a scalpel blade, carefully cut the ligaments connected to pectoral and pelvic muscles. The clavicle and pelvis are easily separated from the plastron by cutting the ligaments and cartilage close to the inner face of the plastron (Fig. **36**). In marine turtles, this marginal bridge of a cartilaginous structure can be easily incised with a knife blade, by inserting the blade laterally into the infra-marginal scales (Jacobson, 2007). The incision should be continued cranially to provide access to the organs of the neck. An incision is made on the margins of the lower jaw and then extended to the oropharynx to provide access to the tongue, glottis, and proximal trachea. The tongue should be lifted and externalized (Munson, 2000; Survey, 2000; Flint *et al.*, 2009; Farris *et al.*, 2014).

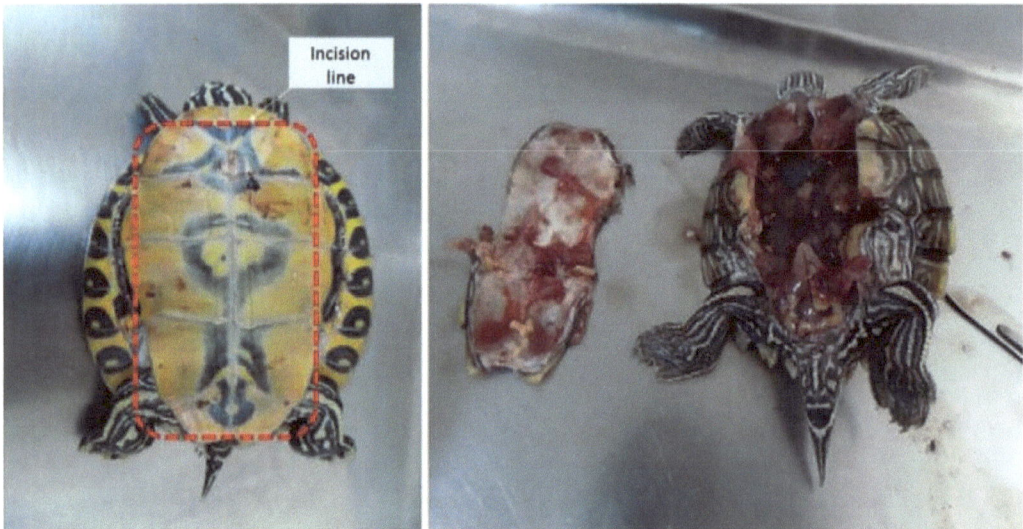

Fig. (36). Representation of the incision line in a turtle.

The endocrine organs present in the neck region should be the first to be observed. The thymus is an even, yellowish-white organ and is located adjacent to the carotid artery or subclavian artery (Jacobson, 2007). The thyroid is a single organ located at the base of the right and left aorta (Mader, 2006). The parathyroid can be difficult to identify and normally is located adjacent to or between the lobes of the thymus (Jacobson, 2007).

The anterior and posterior limbs should be removed in order to provide better access to the celomic cavity. The limbs should be disarticulated at the bones articulation near the shoulder and pelvis and musculature associated with them cut (Jacobson, 2007). After exposing the coelomic cavity, all the organs should be observed *in situ*. The presence of free fluid, fibrous or purulent material in the coelomic cavity should be recorded (Mader, 2006).

Observe the shape, size and colouration of the hearth *in situ*. The pericardial sac should be opened to observe the presence of fluid. In small quantities, it should be considered physiological. When opening the pericardium, a *fibrous gubernaculum cordis*, which connects the apex of the heart to the sac of the pericardium, can observe (Munson, 2000; Survey, 2000; Flint *et al.*, 2009; Farris *et al.*, 2014). The heart should be removed, without damaging the great vessels. The hearth is constituted by three chambers (two atria and one ventricle) (Jacobson, 2007). The heart should be palpated and visually examined for alterations at its shape, colouration, or lesions such as trauma, masses, or foreign bodies (Mader, 2006). Subsequently, the chambers should be sectioned to observe the myocardium and the endocardium. In marine turtles, the pulmonary artery contains sphincters on the surface of the endothelium that resemble an accordion (Jacobson, 2007). Large vessels should be inspected to observe if there is the presence of atherosclerosis or dystrophic mineralization (Mader, 2006).

Next, the gastrointestinal system and the attached glands should be observed. The liver is bilobed and in normal situations has a homogenous colouration of dark brown tone and abundant melanic pigmentation. It can have a paler colour in case of lipidosis and darker in situations of hypertrophy. The liver should be observed and palpated to identify abnormal structures. It should be separated from the stomach by cutting the ligaments (Mader, 2006). The liver should be sectioned for the identification of possible parasites, areas of necrosis, abscesses or alteration of the walls of the bile ducts and blood vessels (Mader, 2006). The gallbladder has a greenish colour and is usually located in the right lobe of the liver. It should be sectioned to observe the mucosa and the contents, observing the presence of any obstructions or calculations (Munson, 2000; Survey, 2000; Flint *et al.*, 2009; Farris *et al.*, 2014).

The gastrointestinal tract should be removed in a single block through a cut in the proximal oesophagus and large intestine (Mader, 2006), as close to the cloaca as possible. Thereafter, some traction and blunt dissection of the ligaments should be performed in order to release it from the carcass. In most species, the oesophagus has an abrupt curvature at the base of the neck before it is inserted into the stomach. In sea turtles, the lumen of the oesophagus contains long, caudally directed spicules (Fig. **37**) (Jacobson, 2007). With the aid of scissors, the oesophagus, stomach and intestine should be cross-sectioned to allow the observation of mucosa of the lumen, to detect the presence of lesions such as ulcers, trauma, masses, or other lesions. As described before, the content should be evaluated and, if necessary, collected for complementary exams (Munson, 2000; Survey, 2000; Flint *et al.*, 2009; Farris *et al.*, 2014).

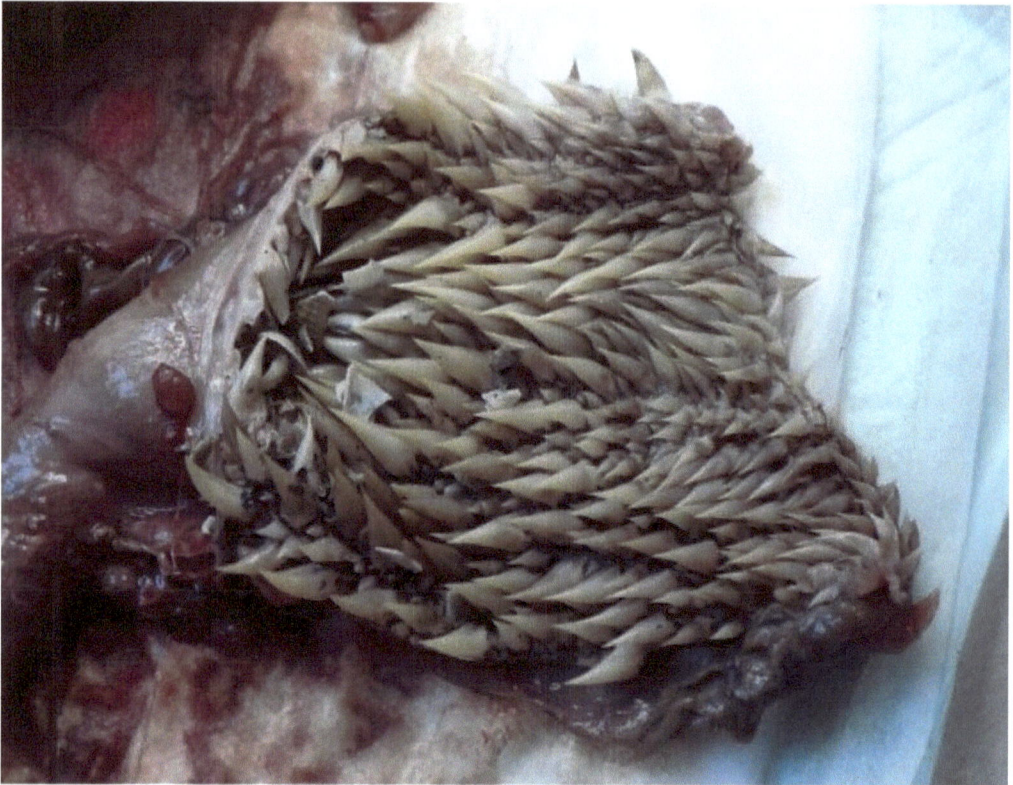

Fig. (37). Detail in a sea turtle (*Caretta caretta*) of the lumen of the oesophagus showing the long, caudally directed spicules.

The spleen is an ovoid organ with a light pink colour in normal situations. It is located near the pyloric sphincter and duodenum, inserted in the mesenteric. It should be observed as fresh as possible as it degrades quickly (Mader, 2006).

The respiratory system should be then observed. The examination should begin by the pharynx and trachea. This one has complete cartilaginous rings. The lungs are quite large, with multiple chambers and with a single intrapulmonary bronchus. In the chelonians, the lungs adhere to the carapace (Mader, 2006). They should be removed carefully after sectioning the trachea in the more proximal region. The traction of the lungs and trachea should be made using forceps and cutting the connective tissue that connects the lungs to the carapace (Jacobson, 2007). An incision should be made along the trachea with the aid of scissors, in order to observe the presence of possible changes in the mucosa and abnormal contents (liquid, food, foreign bodies, parasites) in the tracheal lumen. The lungs should be carefully palpated to detect the presence of firm, nodular areas or other lesions. Transverse cuts are then made in the lung to detect the presence of parasites or exudates (Munson, 2000; Survey, 2000; Flint *et al.*, 2009; Farris *et al.*, 2014).

The kidneys, gonads and adrenal glands can be found in the caudal region of the lungs. The kidneys are located in the retrocelomic space, next to the plastron. A thin membrane separates the kidneys from the coelomic cavity (Mader, 2006). The kidneys have a homogenous brown coloration, elongated shape and are lobed. They should be examined for the possible presence of lesions (Jacobson, 2007). They have ureters and short vas deferens that are inserted in a urogenital bell, that later opens in the ground of the cloaca. The urine passes retrograde from the urodeo to the urine bladder (Fig. **38**) (Jacobson, 2007).

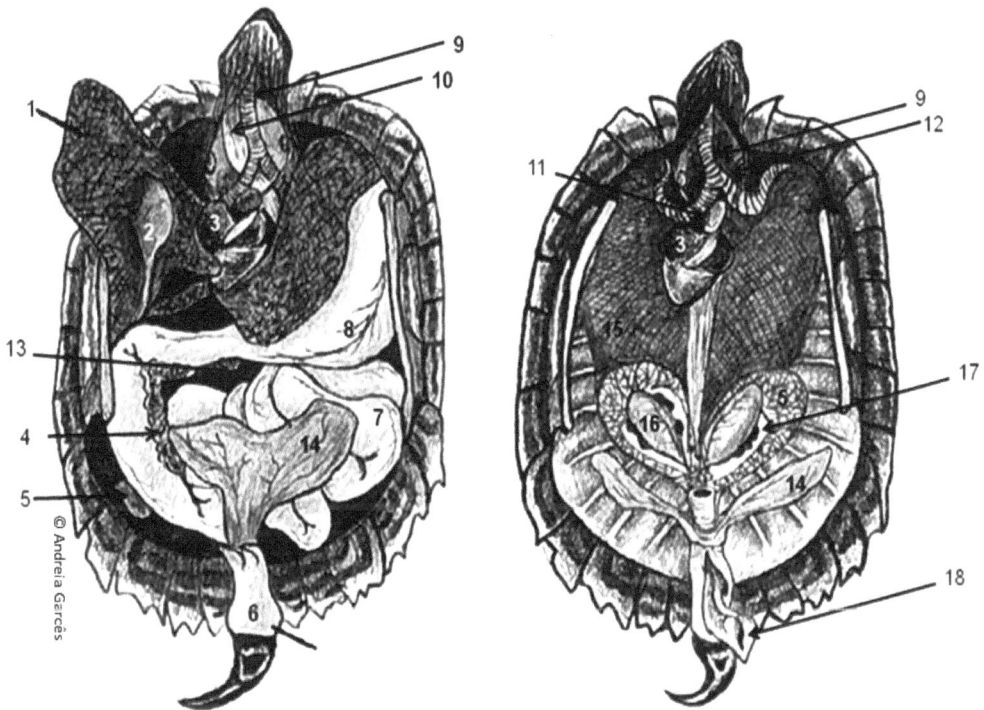

Fig. (38). Schematic representation of the internal organs of a male chelonian (**1**-liver, **2**-gallbladder, **3**-heart, **4**-pancreas, **5** - kidneys, **6**-large intestine, **7**-small intestine, **8**-stomach, **9**-trachea, **10** - oesophagus, **11** - thyroid, **12** - parathyroid, **13** - spleen, **14** - bladder, **15** - lung, **16** - testicles, **17** - adrenal gland).

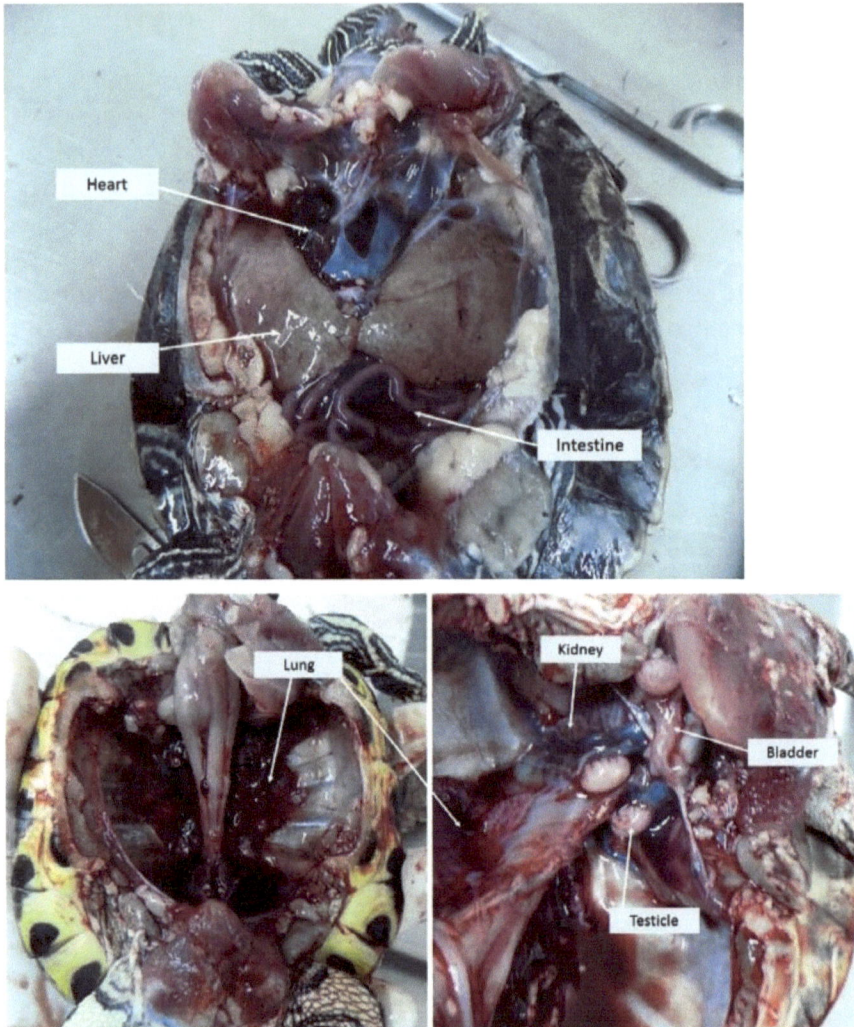

Fig. (39). Internal exam in a turtle.

The urinary bladder is located near the distal colon and the cloaca, in a ventral position to the rectum. This may have two or more lobes. In sea turtles, it may, in addition to urine, contain some mucus, and the mucosa appears rougher with some dark pigmentation. The urinary bladder should be removed and the mucosa should be observed for the presence of any stones and parasites (Mader, 2006). The glands are caudal to the kidneys positioned in the midline. They are elongated in pairs and their colouration may be yellowish-orange. In some species, they are fused (Fig. **39**) (Munson, 2000; Survey, 2000; Flint *et al.*, 2009; Farris *et al.*, 2014).

Females are oviparous. The ovaries are located cranially to the kidney and contain multiple follicles of different sizes. The oviducts flow into the uterus and the vagina, and finally into the cloaca, where the eggs are expelled (Fig. **40**) (Jacobson, 2007).

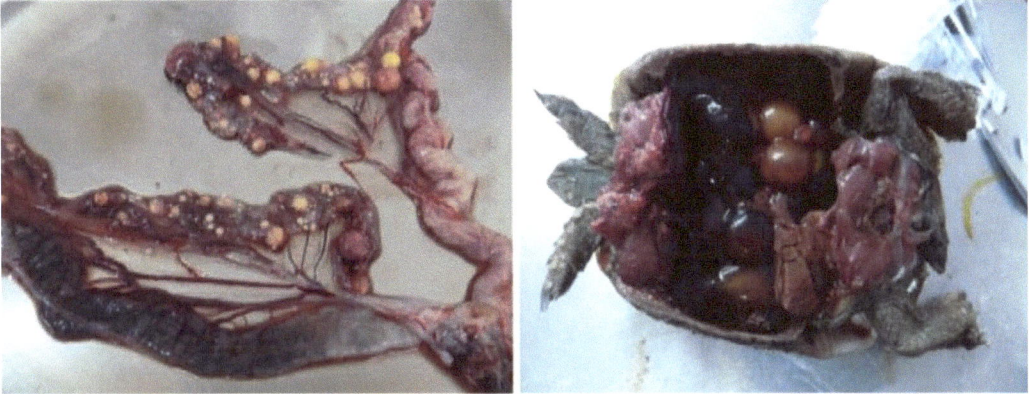

Fig. (40). Genital system in female turtles.

In males, the testes have an elongated shape, yellowish colour and are located in the cranioventral pole of the kidney (Mader, 2006). In the ventral region of the tail the penis is located. These organs should be removed and examined to observe if there are the presence of any changes (Jacobson, 2007).

At this stage, we should observe the presence of deposits of fat under the carapace. They possess a green to brownish colour in healthy animals. If they present a gelatinous and aqueous appearance is indicative of a cachectic animal (Munson, 2000; Survey, 2000; Flint *et al.*, 2009; Farris *et al.*, 2014).

The joints of the limbs should be sectioned to observe the presence of arthritis or gout, and among other lesions (Mader, 2006).

To observe the brain, the animal should be decapitated at the level of the atlas. In small animals, (skull less than 2 cm), the head can be placed directly in fixative (formalin) and then decalcified to section the brain. In larger animals, depending on their size, the skull should be open with the help of a pair of scissors, or a bone saw.

In sea turtles, it is possible to observe the salt gland located next to the eyes. This organ has a firm, lobular form and a pink to light brown colouration (Mader, 2006).

The vertebrae in these animals are fused to the carapace (Jacobson, 2007). To examine the spinal cord, in small animals, the segments of vertebrae should be sectioned and fixated in formalin to decalcify them. In larger animals, it should be accessed by laminectomy (Munson, 2000; Survey, 2000; Flint *et al.*, 2009; Farris *et al.*, 2014).

REFERENCES

Brooks, JW (2018) *Veterinary Forensic Pathology.* Springer International Publishing, NY, USA.

Chai, N *Overview of Reptile medicine.* https://outdoorvets.files.wordpress.com/2010/03/norin-cha--_overview-of-reptile-medicine.pdf

Flint, M (2009) *Necropsies of Reptiles: Recommendations and Techniques for examining Invasive Species.* IFAS Extension University of Florida, Florida.

Garcês, A & Pires, I (2017) *Manual de Técnicas de Necrópsia em Animais Selvagens.* Arteology, Porto.

Girling, S, Raiti, P (2004). *BSAVA Manual of Reptiles.* British Small Animal Veterinary Association, UK.

Huchzermeyer, FW (2009). *Crocodiles: biology, husbandry and diseases.* CABI Publishing, UK.

Jacobson, ER (2007). *Infectious diseases and pathology of reptiles : color atlas and text.* CRC/Taylor & Francis, USA.
[http://dx.doi.org/10.1201/9781420004038]

King, JM (2013) *The necropsy book: A Guide for Veterinary Students, Residents, Clinicians, Pathologists, and Biological Researchers.*College of Veterinary Medicine, Cornell University, NY, USA..

King, JM (2014) *The necropsy book: A Guide for Veterinary Students, by The Necropsy Book.* Charles Louis David DVM Foundation Publisher, Ithaca.

Mader, D (2006). *Reptile Medicine and Surgery,* 2nd ed, Missour: Saunders

Meredith, A, Redro, S (2002). *BSAVA Manual of Exotic Pets.* British Small Animal Veterinary Association, UK.

Mullineaux, E, Best, D & Cooper, JE (2003) *BSAVA manual of wildlife casualties.* British Small Animal Veterinary Association, UK.

Munson, L (2000) *Necropsy of Wild Animals.* Wildlife Health Center, School of Veterinary Medicine, USA.

O'Malley, B (2005) Avian anatomy and physiology. *Clinical: anatomy and Physiology of Exotic Species.* Elsevier, London.

Somvanshi, R & Rao, JR (2009) *Necropsy techniques and necropsy conference manual.* Veterinary Research Institute, India.

Work, TM (2000) *Manual de necropsia de tortugas marinas para biologos en refugios o areas remotas.*U. S. Geological Survey, National Wildlife Health Center Hawaii Field Station, Hawaii, USA.

Terio, KA, Macloose, D, St. Leger, J (2018). *Pathology of Wildlife and Zoo Animals.* Academic Press, UK.

Viner, TC & Kagan, RA (2018) Forensic Wildlife Pathology. *Pathology of Wildlife and Zoo Animals* Elsevier, Oxford 21-40.
[http://dx.doi.org/10.1016/B978-0-12-805306-5.00002-X]

Woodford, MH, Keet, DF & Begins, RG (2000) *Post-mortem procedures for wildlife veterinarians and field biologists, Iucn.* Iucn, Paris, France.

CHAPTER 7

Necropsy in Amphibians

Outline: In this chapter, we describe the method of necropsy in wild amphibians, offering some information regarding the different orders and anatomic characteristics of determined species.

Keywords: Amphibians, Animals sentinels, Aquatic animals, Conservation, Larvae, Mortality, Necropsy, Pathology, *Post-Mortem*.

GENERAL CONSIDERATIONS

There are around 7.000 species of amphibians worldwide. These belong to the subclass Lissamphibia. This class is divided into three orders: Urodela (tetrapods with tail, as is the case of salamanders), Anura (short bodies without a tail, where frogs and toads are included) and Gymnophiona (they have no limbs or tail) (Fig. **1**). They inhabit a great variety of ecosystems, being able to be found in all the continents except in Antarctica. It is characteristic of this Class that the beginning of their life as larvae is in the water, passing later by a metamorphosis in which the gills of the larvae transform into lungs. They also use their skin as a secondary respiratory surface. Some species of salamanders and toads do not have lungs, persisting entirely from breathing through the skin. The smallest amphibian in the world is the New Guinea frog (*Paedophryne amanuensis*), which is only 7.7mm long and the largest in the Chinese Salamander (*Andrias davidianus*) with 1.8m long (Fig. **1**) (Wright and Whitaker, 2001; Price *et al.*, 2014).

Many species of frogs produce mucus and release substances that give them a bad taste to get rid of predators. Some species of amphibians have excretory glands of toxins, existing about 200 toxins already identified. These animals usually have gaudy colours, such as black and yellow. This pattern is also imitated by other non-toxic species to deceive predators, so it is necessary to know the species well.

Due to their particularities, they often serve as indicators of the health of their ecosystem. At present, most of the planet's amphibian populations are threatened, with some already extinct due to various factors, such as pollution, destruction of habitats, or outbreaks of infectious diseases.

It is increasingly common to observe animals with malformations in limbs due to environmental factors and pollution. Several infectious agents, such as *Batrachochytrium dendrobatidis* and Iridovirus have been responsible for decimating entire populations. For this reason, the *post mortem* examination of amphibians is increasingly essential (Wright and Whitaker, 2001).

Its decomposition process is accelerated, not only due to their habitat type (tropical, humidity) but also because of its rapid metabolism, especially in smaller animals. Therefore, as soon as possible, the corpse should be collected and refrigerated (about 4-10°C) before performing the necropsy should be accomplished no later than 6 hours after death. In larger animals, it may be useful to dip them in ice water or make an incision in the abdomen to accelerate the cooling process. If it is not possible to carry out the necropsy on time, the corpse's fixation in formaldehyde must be considered. In this situation, the coelomic cavity should be opened along the midline to allow rapid fixation of the internal organs. This technique works best on small individuals (Mader, 2006; King *et al.*, 2014).

Fig. (1). Examples of amphibians belonging to the Order Urodela and order Anura.

EXTERNAL EXAM

The body condition of the animal should be evaluated and recorded (Fig. **2**). The presence of thin animals is frequent and may result from infectious diseases or, in the case of captivity, due to malnutrition or competition with other individuals in the habitat. The skin is fragile, moist, and with abundantly glandular areas, that can have smooth or have a sandy appearance. The presence of discolourations increased mucus production, nodules, ulcerations or trauma should be observed. Where possible, swabs of the skin should be performed (Fig. **2**).

The oral cavity, eyes, ears, nostrils and cloaca should also be examined. In the Anuria Order, the nostrils are quite small and are located cranially at the tip of the muzzle.

The eyes are quite large and have a nictitating membrane to protect them. Posterior to the eye, they have a circular tympanic membrane representing the outer ear. This membrane allows distinguishing sexes since in females is the same size as the eye, while in males is much larger (King *et al.*, 2014).

In the case of some species of the Order Caudata, the larvae have three pairs of gills located posterior to the head in the lateral zone. Your mouth is large, small eyes and nostrils communicate with the oral cavity (Wright and Whitaker, 2001).

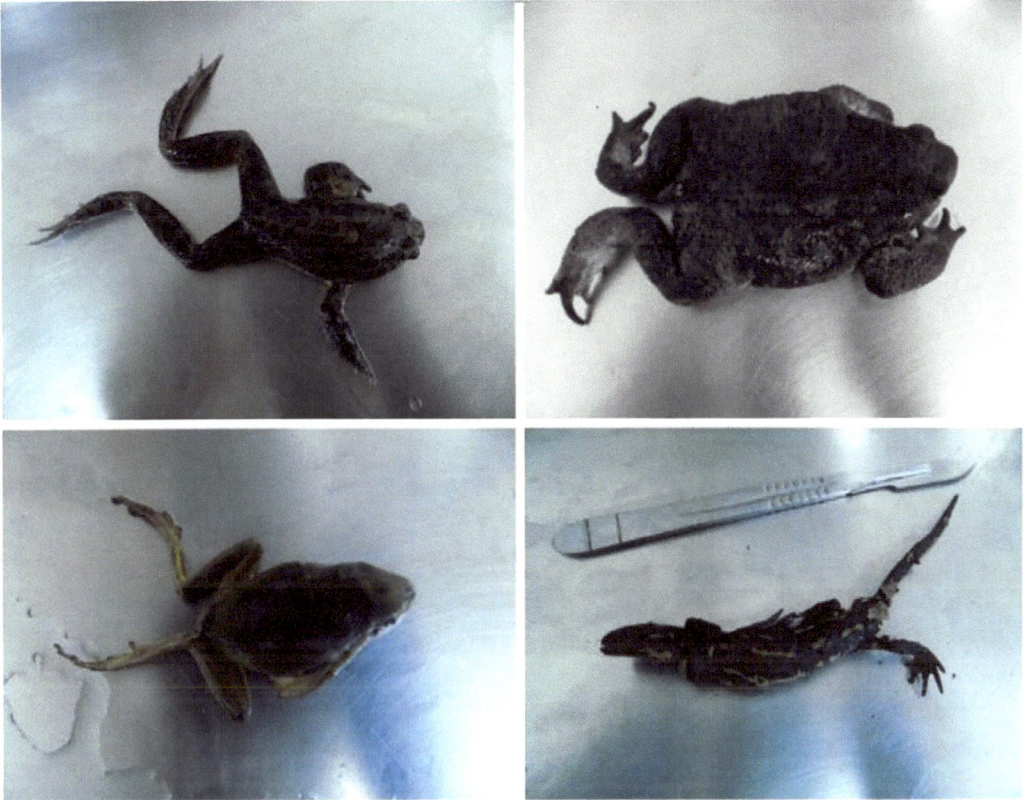

Fig. (2). Examples of some species of amphibians.

In the oral cavity, observe the teeth. All orders have peduncular teeth, which are curved toward the pharynx to push the prey to the stomach. Teeth are being replaced throughout life. Caecilians, salamanders and some anurans have one or two rows of maxillary and mandibular teeth.

The tongue of most anurans and salamanders can be extended beyond the mouth to capture food, much like chameleons. Exceptions are the Caecilians who have fixed tongues and some frog species that do not have a tongue (Figs. **3-4**).

Examination of the oral cavity could detect traumatic lesions, parasites or fungi. Swabs/cytology's of the oral cavity should be performed to observe whether protozoa, such as *Trichodina* and *Piscinoodinium* or fungi are present (Wright and Whitaker, 2001; Garcês and Pires, 2017).

Some species have a claw-like structure in the hind limbs as can be observed in the *Xenopus laevis* frog species and the salamanders of the genus *Onchydactylus spp* (Wright and Whitaker, 2001).

Fig. (3). External exam in a *Lissotriton boscai*.

INTERNAL EXAM

To the internal examination, the animal should be placed in a ventrodorsal position. An incision is made through the midline, from the intermandibular region to the anus. The skin is removed ventrally to expose the abdominal muscles (Figs. **5-8**). If the present subcutaneous fluid is, it should be collected for cytology and culture. Next, an incision is made in the abdominal wall through the midline to access the coelomic cavity. It should be noted that the amphibian muscle is paler than in other vertebrates (Wright and Whitaker, 2001).

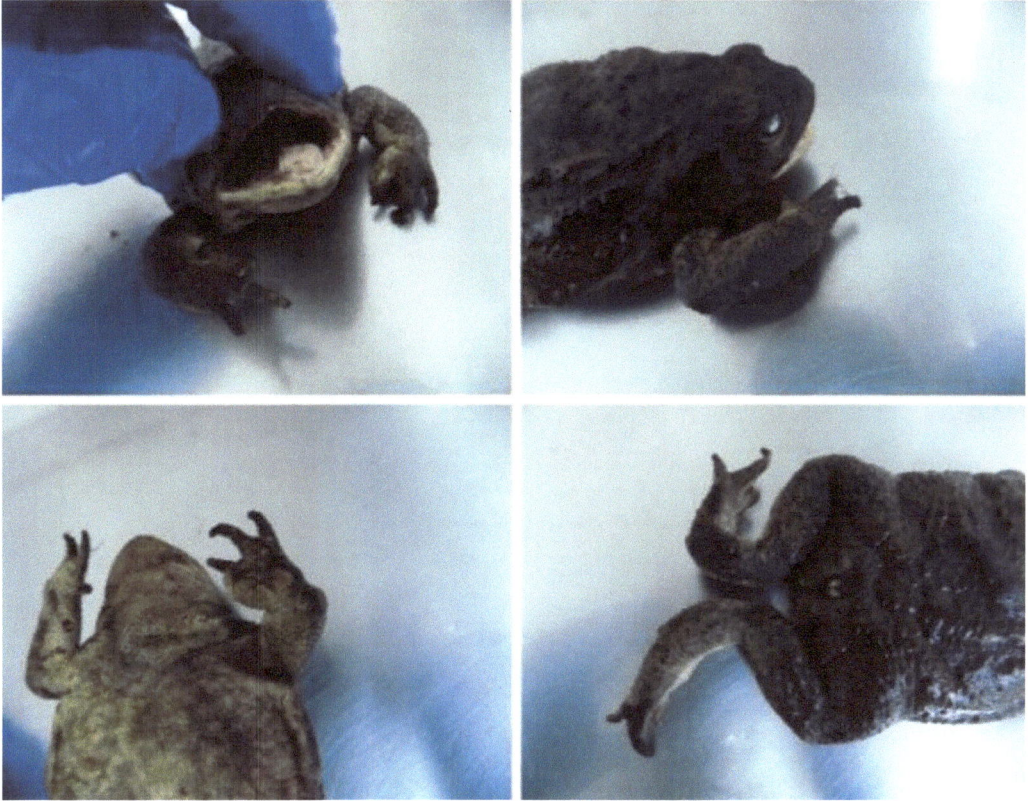

Fig. (4). External exam in a *Bufo bufo.*

Fig. (5). Acess to the coelomic cavity in an individual of the order urodela (**1**-liver, **2**-stomach, **3**-ovary, **4**-intestine, **5**-bladder, **6**-oviduct, **7**-lung).

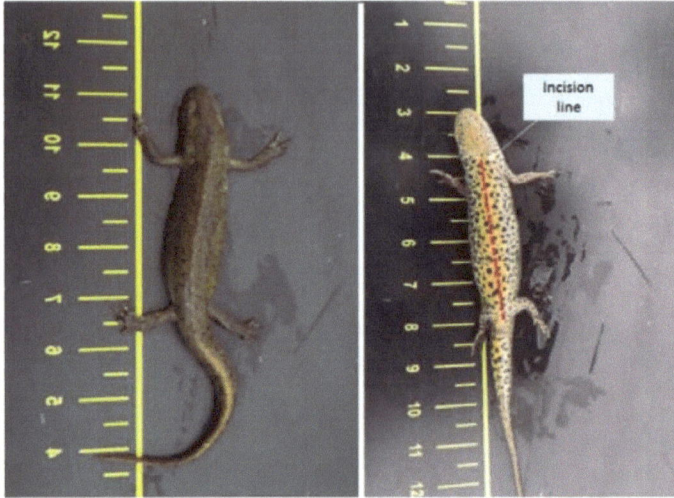

Fig. (6). Representation of the incision line in the Order Urodela.

If present, the amount of free fluid in the coelomic cavity should be quantified and collected for further examination. Without damaging the viscera, the heart is exposed, by cutting the sternum. All organs *in situ* are observed at this stage to identify changes in organ position or presence of lesions such as nodules, enlargements, fibrin accumulation, or exudates (Terio, McAloose and Leger, 2018).

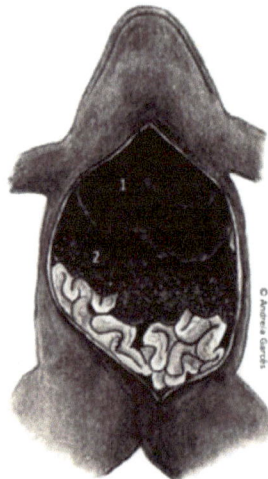

Fig. (7). Acess to the coelomic cavity in an animal of the Order Anura (**1**-liver, **2**-ovary, **3**-intestine).

After the incision, the fat bodies should be observed to help to assess its body condition (Figs. **7,8**). Fat bodies that are enlarged and present throughout the coelomic cavity indicate that it is an obese animal, whereas small fat bodies or the apparent non-existence of these suggest anorexia (Wright and Whitaker, 2001).

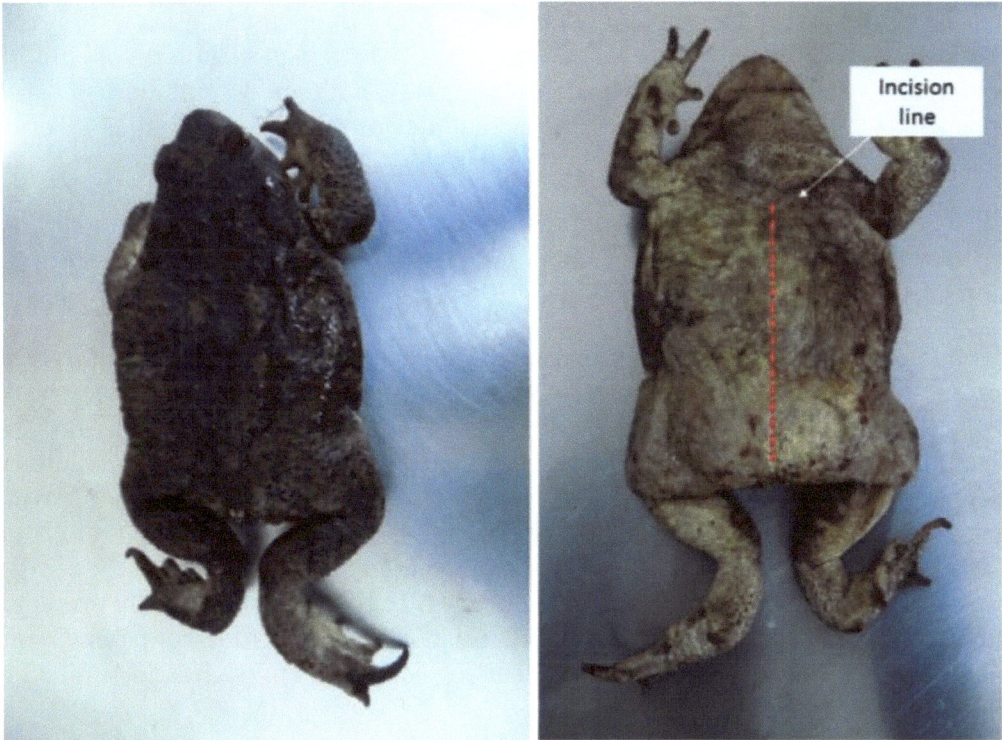

Fig. (8). Representation of the incision line in the Order Anura.

The internal organs should be removed in the block starting by the separation of the tongue, pharynx and glottis from the surrounding tissue. The tongue is then drawn in the caudal direction, and the remaining organs are separated from the adjacent tissues to which they are attached in a block (Terio, McAloose and Leger, 2018).

The heart of amphibians has three chambers: two atria and one ventricle. The interatrial septum is fenestrated in the caecilians and most salamanders while in the anurans, it is complete. In most cases, the heart can be fixed as a whole in formalin. In larger animals, an incision is made in the atrium and ventricle. Examine the inside to detect if there is the presence of thrombi or nodules. The apex of the heart lies near the largest lobe of the liver. Adhesions present between the two are common, often resulting from bacterial pericarditis (Wright and Whitaker, 2001).

The lungs, when present, are simple saccular structures. Cartilage rings support the trachea, and its length varies with the species, being usually short, branching into the main bronchi. Amphibians do not have a diaphragm, the gas exchanges being dependent on the coordinated movement of their axial and appendicular muscles. Often the lungs may be hyperinflated, which is common because it is a response to stress situations, used as a defence mechanism. The lung should be examined to evaluate changes in colour, masses, nodules or granules, and the presence of exudates or parasites (*e.g.* nematodes of the genus Rhabdias) (Terio, McAloose and Leger, 2018).

When present, gills should also be examined. They have significant variability depending on the species, environment and stage of the life cycle in which they are found.

The liver is then separated from the remaining viscera. In the Anuria, the liver is bilobed. In the Caudados slightly elongated and the Caecialians are very elongated. The gallbladder is present. Its normal colouration is reddish-brown. Changes in pale or yellowish colouration may be inferred to extramedullary haematopoiesis or lipidosis. The black colour is commonly associated with melanchrophagous hyperplasia in cases of antigenic stimulation, starvation or weakness. These changes in colour are also physiological when related to specific stages of the cycle or the year season. It is also observed its size and the presence of bruising or trauma. Several incisions are made to examine the liver parenchyma and detect if present, granulomas, necrosis or neoplasms (Wright and Whitaker, 2001).

The spleen has a spherical shape with dark red colouration and is located adjacent to the stomach and duodenum and observed to be evaluated for changes in size, colour or conformation (Figs. **9, 10**) (Wright and Whitaker, 2001).

The gastrointestinal tract in these animals is usually short and straightforward. Its surface should be observed to record the presence of any changes in colour, masses, adhesions, trauma or fibrin exudates. It should be opened along its length with a pair of scissors to observe the lumen and to detect the presence of areas of discolouration, ulcers or parasites, among other lesions (Terio, McAloose and Leger, 2018).

The kidneys are elongated and usually have a reddish-brown colour. These should be removed and examined for changes in size, colour, presence of granulomas, or neoplasms.

The bladder is a thin ventral structure to the large intestine, bilobed in many Caecilians. It should be palpated and open to observe lesions on its lumen, such as calculations (found in uricotelic species such as the *Phyllomedusa* sp.), parasites, haemorrhages or nodular formations (Wright and Whitaker, 2001) (Wright and Whitaker, 2001).

Along with the urinary system, observe the genital system. Amphibians have a pair of testicles or ovaries that lie within the coelomic cavity. Changes in consistency, colour, size, presence of masses or cysts should be recorded (Figs. **11-12**).

To access the brain, we begin by separating the skull from the spine, exposing the *forumen magnum*. Then an incision is made on the skin, and it is removed to expose the skull. Then depending on the size of the animal, the cranial box can be opened with scissors or a bone saw. In small animals (skull less than 2 cm), the head can be placed directly on formalin and then decalcified to section at the brain (Wright and Whitaker, 2001).

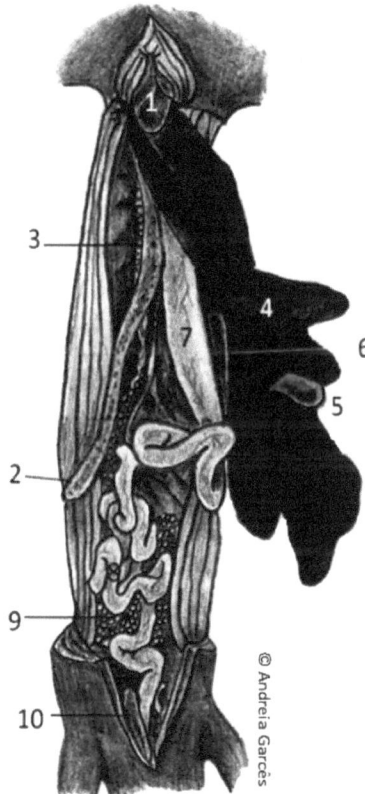

Fig. (9). Coelomic cavity in Urodela species (**1**-heart, **2**-lung, **3**-oviduct, **4**-liver, **5**-vesicle, **6**-spleen, **7**-stomach, **8**-intestine, **9**-ovary, **10**-bladder).

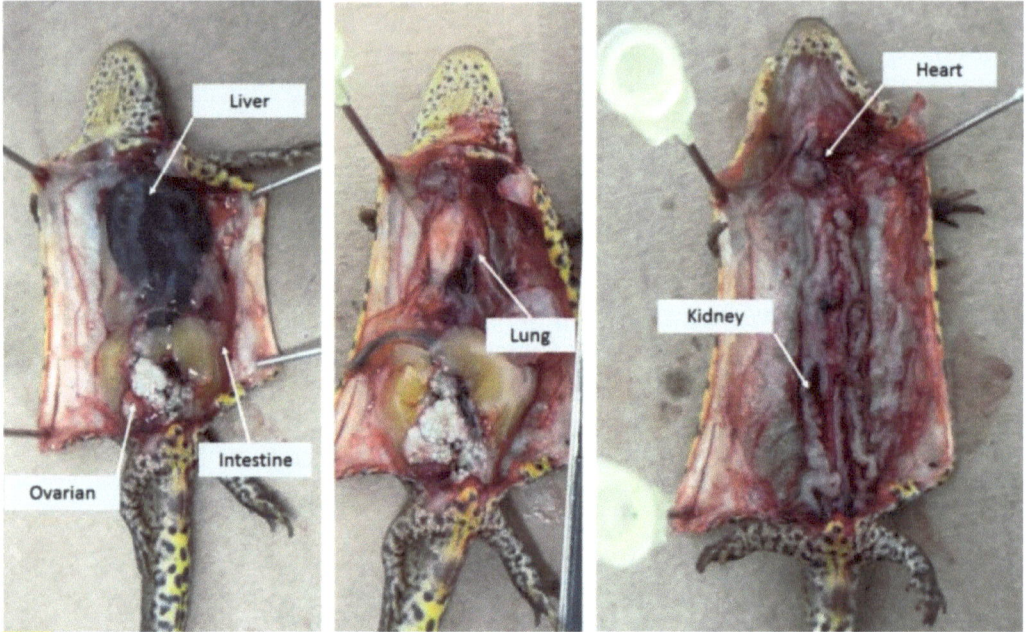

Fig. (10). Internal exam in a *Lissotriton boscai.*

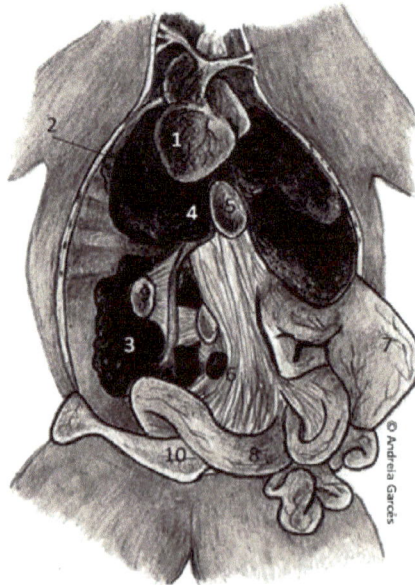

Fig. (11). Opening of the coelomic cavity in an individual of the species of the Order Anura (**1**-heart, **2**-lung, **3**-kidney, **4**-liver, **5**-vesicula, **6**-spleen, **7**-stomach, **8**-intestine, **9**-testiculum, **10**-bladder).

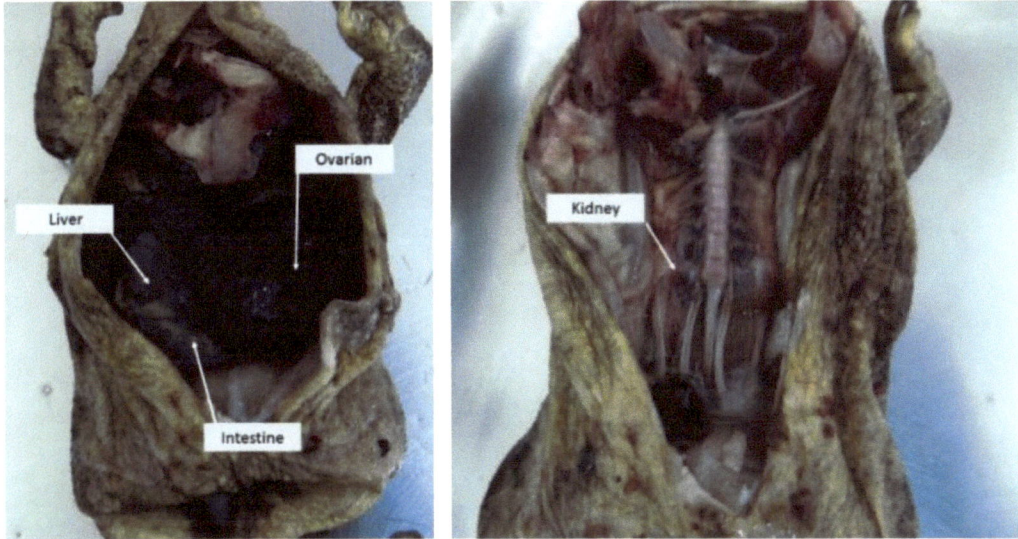

Fig. (12). Internal exam in a *Bufo bufo.*

REFERENCES

Garcês, A & Pires, I (2017) *Manual de Técnicas de Necrópsia em Animais Selvagens,.* Arteology, Porto.

King, JM (2014) *The necropsy book: A Guide for Veterinary Students, by The Necropsy Book* Charles Louis David DVM Foundation Publisher, Ithaca.

Mader, D (2006). Reptile Medicine and Surgery, 2nd ed, Missour: Saunders,

Price, SJ, Garner, TW, Nichols, RA, Balloux, F, Ayres, C, Mora-Cabello de Alba, A & Bosch, J (2014) Collapse of amphibian communities due to an introduced Ranavirus. *Curr Biol,* 24, 2586-91. [http://dx.doi.org/10.1016/j.cub.2014.09.028] [PMID: 25438946]

Terio, KA, McAloose, D, St. Leger, J *Pathology of Wildlife and Zoo Animals.* 1st ed (2018) Academic Press, UK.

Wright, K N & Whitaker, B R (2001) *Amphibian Medicine and Captive Husbandry.* Krieger Publication, USA.

CHAPTER 8

Necropsy in Invertebrates

Outline: In this chapter, we describe the method of necropsy in invertebrates, offering some information regarding the different orders and anatomic characteristics of certain species.

Keywords: Animals sentinels, Conservation, Insects, Invertebrates, Molluscs, Mortality, Necropsy, Pathology, *Post-Mortem.*

GENERAL CONSIDERATIONS

The necropsy of invertebrates is important because it allows increasing the knowledge about pathological alterations and the normal tissues in these animals, often ignored (Fig. **1**).

The greatest difficulty during necropsy in invertebrates is to be certain of their death. For this reason, whenever possible, euthanasia should be performed in moribund animals. In addition, the tissues of invertebrates deteriorate much faster than vertebrates. In most cases, as it is often not possible to determine the exact time of death of the animal, the *post-mortem* procedure should be done as soon as possible (Fig. **1**) (Meredith and Redro, 2002).

The equipment is the same as that used for a vertebrate necropsy. Besides, smaller animals need to have microscopic dissecting material (Lewbart, 2012).

EXTERNAL EXAM

The exoskeleton of these animals should be carefully observed to detect lesions that could be related to possible diseases. In squares, the first structure to examine is the carapace where can be observed traumatic lesions, parasites, pustules, blisters or malformations (Figs. **2,3**) (Meredith and Redro, 2002).

Fig. (1). Examples of invertebrate animals.

In cephalopods, ulcerations on the skin, oedema or presence of excess mucus in the tentacles are common. In equidae, the presence of discolouration, for example, red or black that are indicative of inflammation, loss of peaks or changes in these can be notable. In crustaceans, the external exam is necessary to observe the presence of erosions in the carapace or blackening of the gills (an indicator of fungi) (Lewbart, 2012).

Fig. (2). External habit of a tarantula.

Fig. (3). External habit of a *Danaus plexippus.*

INTERNAL EXAM

In insects, the exoskeleton is made up of chitin. Normally, a section allows to access the internal tissues but, in cases where this layer has a greater thickness, the animal can be placed in descaler solutions, since there is calcium deposited in the cuticle (Meredith and Redro, 2002). In most corals and echinoderms necropsies, decalcification may be necessary. In some crustaceans, the carapace is easily removed, and no decalcification is necessary except the beak of some cephalopods or in the case of small bivalves (Lewbart, 2012).

Fig. (4). Summary of *post-mortem* examination in a *Poecilotheria regalis.*

After the incision, tissue consistency, colour changes, signs of infection, cysts or tumour masses should be observed (Figs. **4 - 6**). In crustaceans, haemolymph staining is observed: red is an indicator of infection by *Aerococcus viridans var. homari* or *paramoeba*, orange of mineral diseases and whitish, in crabs, from infection by *Hematodinium sp.*

Whenever possible, cytology and swabs of organs are recommended.

Attention should be paid to venomous animals such as anemones or tarantulas, so as not to touch the poison. This should be as soon as possible deactivated in formaldehyde (Fig. **4**) (Lewbart, 2012).

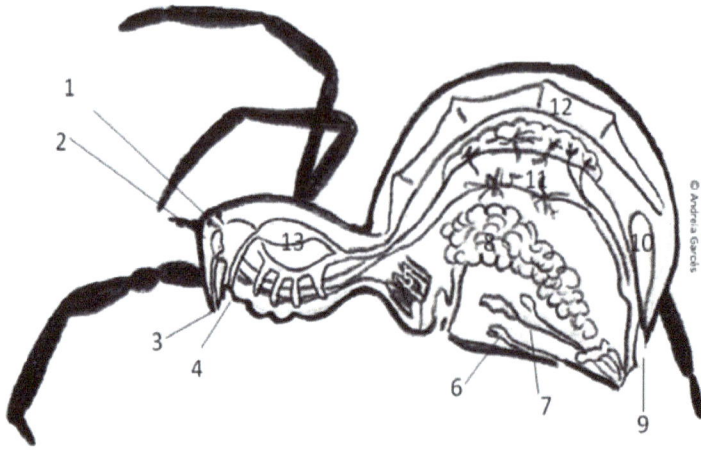

Fig. (5). Anatomical representation of a tarantula (**1**-brain, **2**-gland of venom, **3**-tooth, **4**-mouth, **5**-lung, **6**-trachea, **7**-silk gland, **8**-ovary, **9**-anus, **11**-intestine, **12**-heart, **13**-stomach).

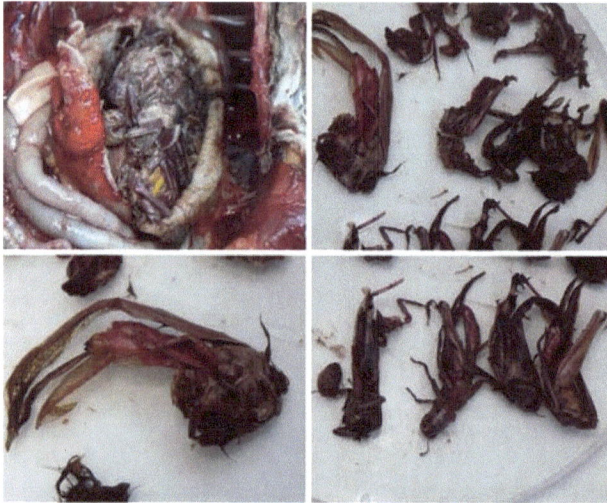

Fig. (6). Locusts removed from the gastric contents of a *Buteo buteo*.

REFERENCES

Lewbart, G (2012). *Invertebrate Medicine.* Wiley-Blackwell, USA.

Meredith, A, Redro, S (2002). *BSAVA Manual of Exotic Pets.* British Small Animal Veterinary Association, UK.

CHAPTER 9

Necropsy in Fish

Outline: In this chapter, we describe the method of necropsy in fish, offering some information regarding the different orders and anatomic characteristics of the determined species.

Keywords: Aquatic Animals, Conservation, Fish, Mortality, Necropsy, Pathology, *Post-Mortem*, Shark.

GENERAL CONSIDERATIONS

There is a wide variety of fish. These animals are aquatic and can inhabit in a wide range of ecosystems, from freshwater to saltwater, and deep-water to shallow water. They have, therefore, a great variety of characteristics that represent the environment in which they inhabit, which is reflected in their great anatomical diversity (Widgoose, 2001). Therefore, anatomical knowledge of the different groups is essential after identifying the species (Figs. **1-10**) (Meredith and Redro, 2002).

Sex determination is often only possible after a *post-mortem* examination, except in cases where there is sexual dimorphism (Widgoose, 2001).

It is essential to know as much as possible about the *ante-mortem* history. In captive animals, we should try to understand what conditions were maintained concerning water temperature parameters, type of feed, CO_2 and O_2 levels, nitrates and nitrites, pH, *etc.* These parameters are fundamental because changes in them often cause the animals to die (Widgoose, 2001).

Post-mortem examination should be as early as possible after death, as these animals go into putrefaction quickly (Roberts, 2012; Smith, 2019).

EXTERNAL EXAM

Immediately after death, the external examination is carried out. All injuries observed, such as signs of trauma or foreign bodies, should be noted, if present. If possible, samples should be collected (Roberts, 2012; Smith, 2019).

Skin, gills, fins, and operculum should be examined for haemorrhages, parasites, ulcers, fungi, microbubbles (an indication of supersaturation), vessel telangasia (indicating large concentrations of ammonia), or swelling (Figs. **2-4**) (Widgoose, 2001).

The grommets shall be lifted and, at this stage, sectioned to access the gills. Dark brownish gills may be indicators of nitrate toxicity, while a pale/pink coloration indicates anemia. In cases of an advanced state of putrefaction, the gills are whitish, which should not be confused with situations of anemia (Roberts, 2012; Smith, 2019).

The fins should be observed to detect signs of corrosion or bleeding that may be signs of bacterial contamination in the water or poor handling of it.

The oral cavity and anus are observed to evaluate the presence of traumatic lesions, edema, or parasites. The eyes should be examined to determine the presence of opacities, hemorrhages, excessive mucus, exophthalmia, endophthalmitis, or buphthalmia. In the anus and genital pore, observe the presence of trauma, oedema, protrusions, and parasites (*e.g.* nematodes), if present. Care should be taken to manipulate some species of the Scorpaenidae family because they are poisonous. The venom glands should be put into formalin to be deactivated (Fig. **1**) (Roberts, 2012; Smith, 2019).

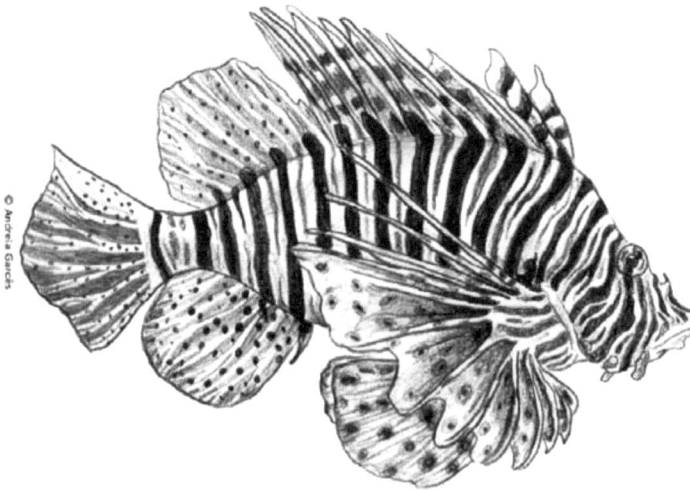

Fig. (1). External exam of a *Pterois-volitans.*

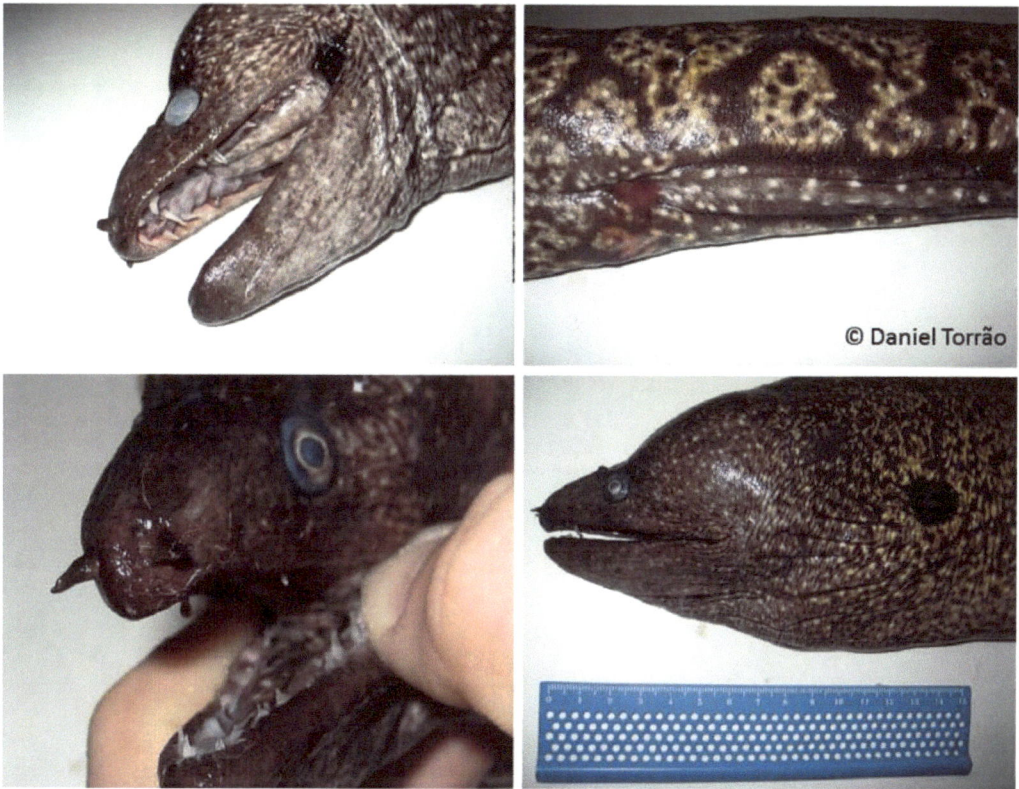

Fig. (2). External exam in a *Muraena helena.*

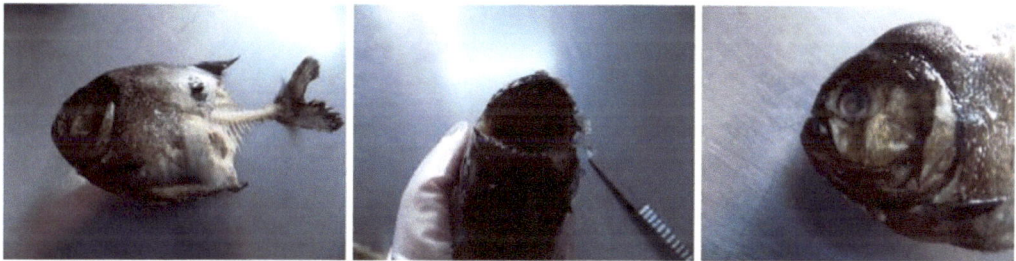

Fig. (3). External exam in a *Pygocentrus nattereri.*

INTERNAL EXAM

The method of the opening varies with the anatomical shape of the fish. In the case of laterally flattened fish such as Goldfish (*Carassius auratus*) or Atlantic salmon (*Salmo salar*) (Widgoose, 2001), the coelomic cavity should be accessed laterally. In the case of fish belonging to the subclass Elasmobranchii, where the rays and sharks are placed, access can be made by the midline of the abdomen (Figs. **5** and **6**). Pleuronectal fish should be examined in lateral decubitus on the depigmented side (Roberts, 2012; Smith, 2019).

Fig. (4). External exam in a *Carassius auratus*.

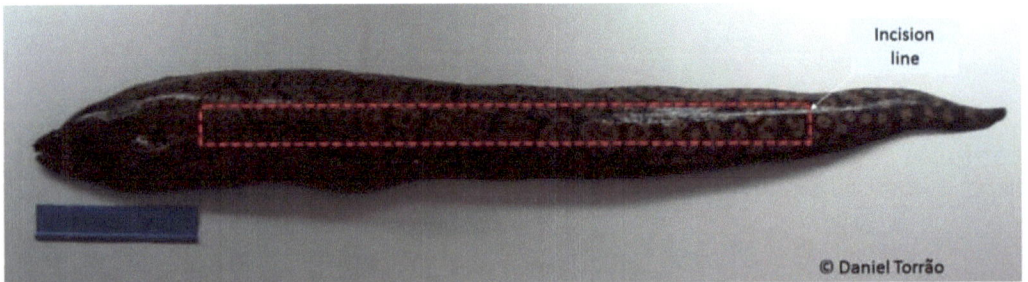

Fig. (5). Incision line in a *Muraena helena*.

Fig. (6). Incision line in a *Carassius auratus.*

In the case of fish where the access is lateral, with very sharp scissors, the incision is started in the anal opening and progressed along an arc in circumference that draws the upper limit of the coelomic cavity until it meets the first end of the first cut (Roberts, 2012; Smith, 2019). In this case, the sectioned portion of the wall is removed. In other fish, the middle line is used to access internal organs. The internal organs should be observed *in situ* (Fig. **7**) (Widgoose, 2001).

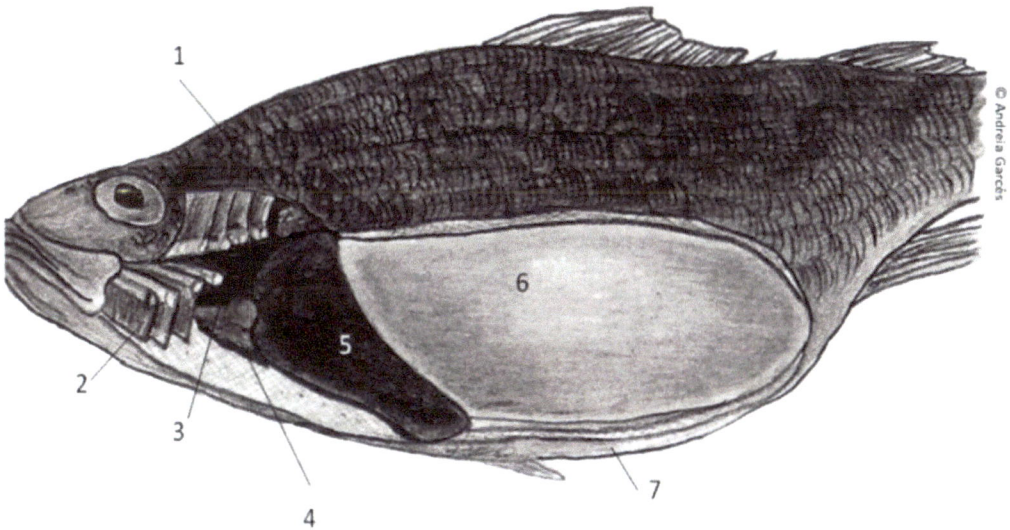

Fig. (7). Removal of the muscular wall with exposure of internal organs (**1**- kidney, **2**-gills, **3**-pharynx, **4**-heart, **5** - liver, **6**-ovary, **7**- intestine).

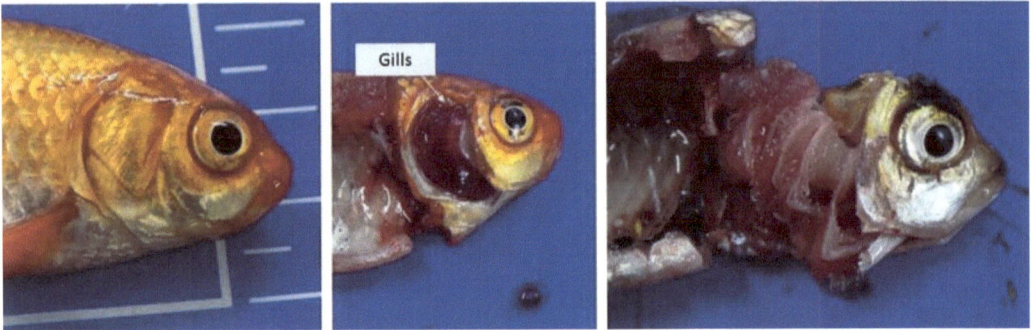

Fig. (8). Removal of the operculum to exposure of the gills.

Then the organs are removed one by one and possible changes are observed. In the swimming bladder, the thickness of the mucosa is observed, with the presence of haemorrhages, necrosis, fungi, and the presence of parasites. (Fig. **10**). In the kidney, the presence of parasites, granulomas, or anatomical abnormalities should be observed (Widgoose, 2001).

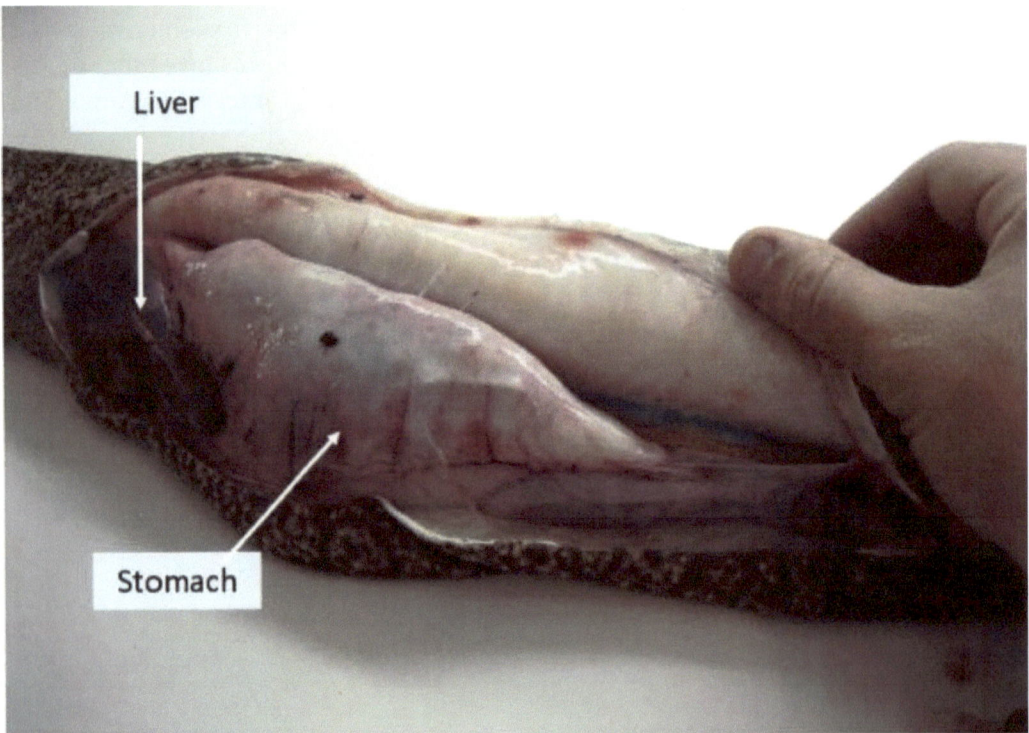

Fig. 9 cont.....

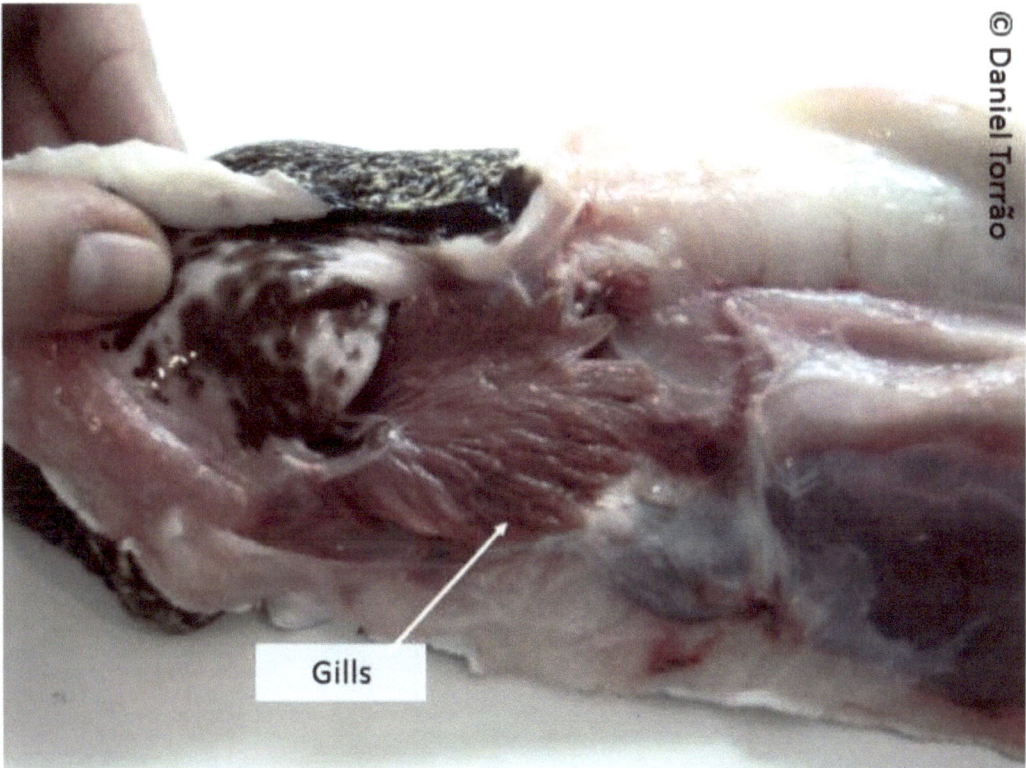

© Daniel Torrão

Fig. (9). Internal exam in a *Muraena helena.*

In the liver, color, consistency, fat content, parasites, or granules are evaluated. The gallbladder is located near or inside the liver and contains a yellowish-green colored liquid typically and may be increased in cases of anorexia (Roberts, 2012; Smith, 2019).

The gonads should be observed for their size, maturity, granulomas, or stable structures (in females, egg degeneration may occur) (Figs. **9-10**) (Widgoose, 2001).

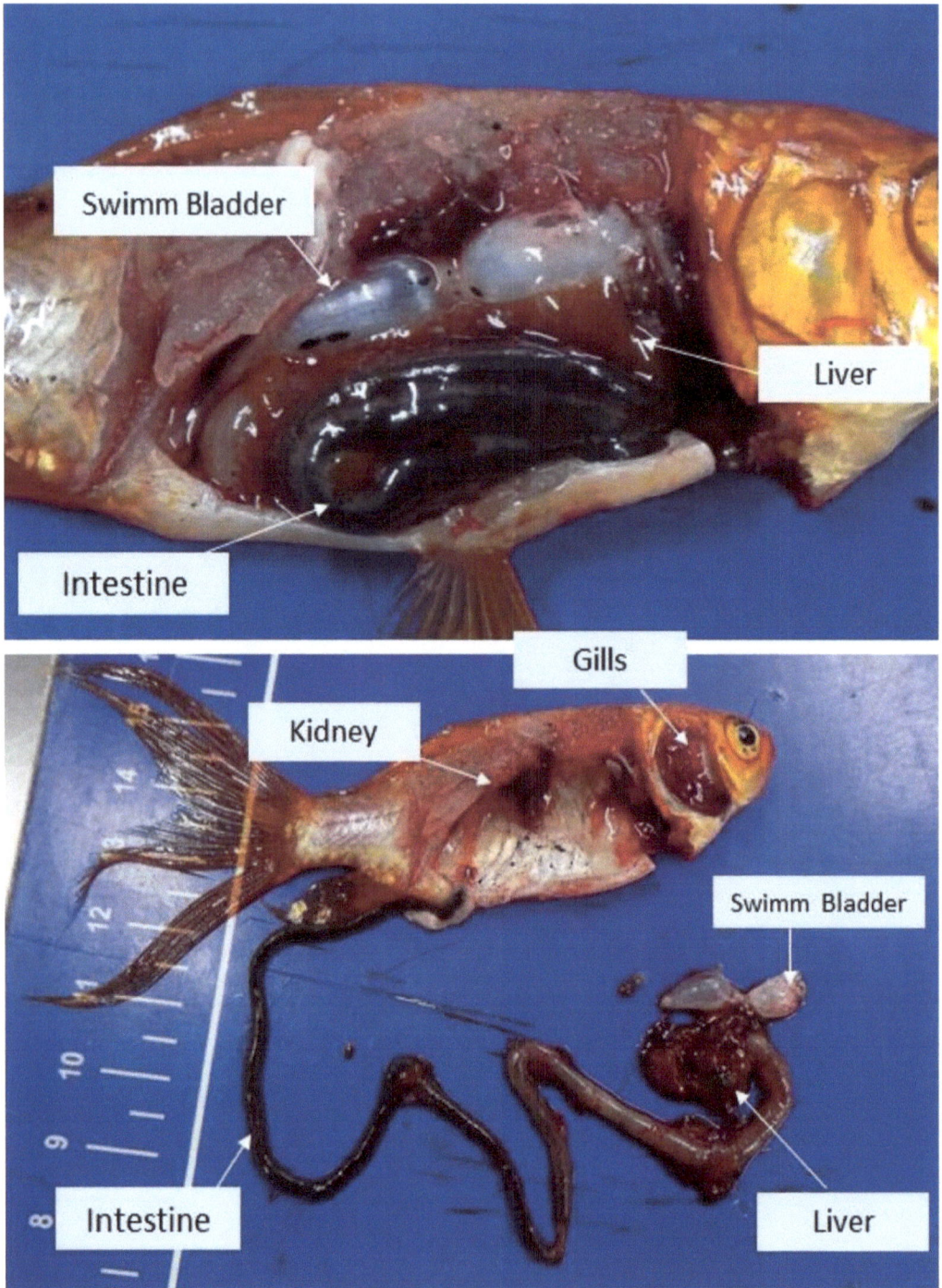

Fig. (10). Internal exam in a *Carassius auratus*.

The stomach is also observed and evaluated for lesions as granules in the stomach. It should be noted that in Koi, goldfish, and other cyprinids, there is no real stomach. In the gut, the presence of food, fluids, hemorrhages, ulcers, and parasites (mainly nematodes) should be recorded (Figs. **11-10**) (Roberts, 2012; Smith, 2019). The brain should be observed to detect the presence of trauma or other lesions (Fig. **12**).

When necessary, tissue samples are removed for histopathology, swabs of organs for microbiological examination, and parasites for identification (Roberts, 2012; Smith, 2019).

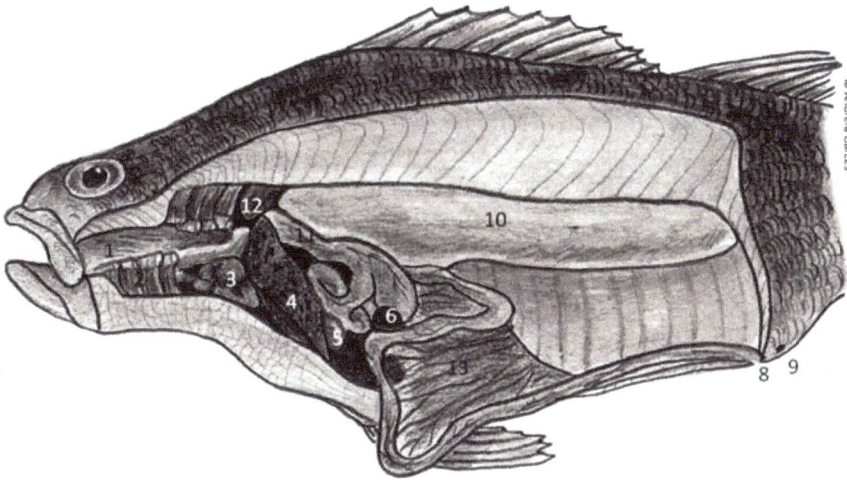

Fig. (11). Scheme of the internal organs of fish (**1**-pharynx, **2**-gills, **3**-heart, **4**-liver, **5**-gallbladder, **6**-spleen, **7**-rectum, **8**-anus, **9**-urogenital opening, **10**-swimming bladder, **11**- kidney, **13**-gut).

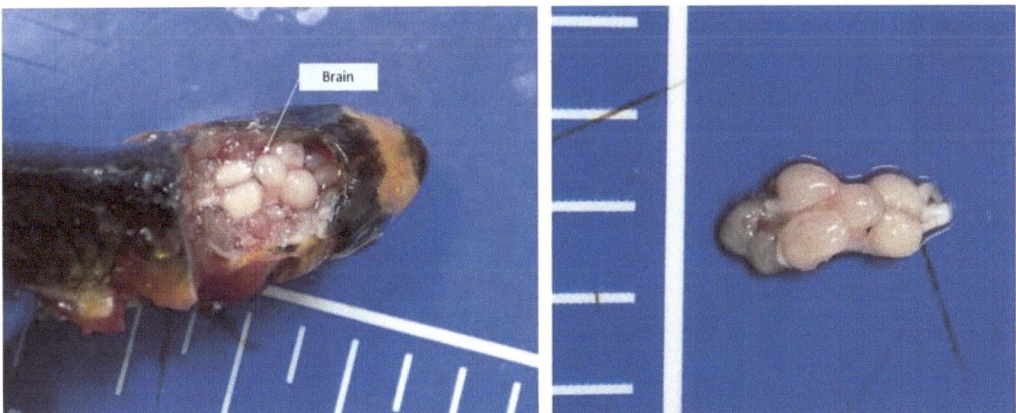

Fig. (12). Exposure of the brain in a *Carassius auratus.*

REFERENCES

Meredith, A, Redro, S (2002). *BSAVA Manual of Exotic Pets.* British Small Animal Veterinary Association, UK.

Roberts, R (2012). *Fish pathology.* Blackwell Publishing Ltd, USA.
[http://dx.doi.org/10.1002/9781118222942]

Smith, S (2019). *Fish Diseases and Medicine.* CRC Press Taylor & Francis Group, Boca Raton.
[http://dx.doi.org/10.1201/9780429195259]

Widgoose, WH (2001). *BSAVA Manual of Ornamental Fish.* BSAVA, London.

CHAPTER 10

Necropsy in Eggs

Outline: In this chapter, we describe the method of necropsy in eggs of wild birds, in wildlife conservation. The authors also describe the necessary equipment, the difficulties presented, and the samples that can be obtained from this procedure.

Keywords: Amphibians, Animals sentinels, Birds, Conservation, Egg, Fish, Mortality, Necropsy, Pathology, *Post-Mortem*, Reptiles.

GENERAL CONSIDERATIONS

Before starting the procedure, it is essential to know the species to which the egg belongs, the history of the progenitors, and data related to the incubation period.

The primary information to be given about incubation includes the date of laying, whether it was artificial or by parent incubation, disturbances during the incubation period, and the duration of incubation till fetal death occurred (Chitty and Lierz, 2008; King *et al.*, 2014).

In the case of artificial incubation, the type of incubator used must be known. It is important to have access to the data on temperature, humidity, light, and ventilation. The storage period and conditions before initiation of incubation should be ascertained (Fig. **1**) (Schmidt and Reavill, 2003). Access to test results that may have been carried out previously and that detect certain diseases, such as candidiasis or salmonellosis, should be available. Also important is the data on the evolution of egg weight over time, data on nests of the same batch, the percentage of eggs that did not hatch, among others (Coles, 2007).

Due to their characteristics, the eggs enter rapidly into autolysis, therefore, necropsy should be performed as soon as possible, or the egg should be refrigerated (King *et al.*, 2013).

Andreia Garcês & Isabel Pires

Fig. (1). Infertile eggs.

EXTERNAL EXAM

The eggs must be weighted and measured for length and diameter (*e.g.* eggs with small diameter may be indicative of salpingitis). The surface must be observed, and the presence of discoloration, stains, fractures, perforation, dirt, or presence of identification numbers should all be recorded (Figs. **2** and **3**). If evidence of lacings/pecks is observed on the bark, it should be assessed whether the internal membranes are visible (Kinne, 2015).

The egg is then placed on the light to observe whether it is fertilized or not (Schmidt and Reavill, 2003; Butcher and Miles, 2015; Garcês and Pires, 2017).

During this period, a scheme of what has been observed is performed, regarding the position and size of the air space and location of the embryo, if present, and other lesions.

All theses procedures and lesions must be photographed with a scale.

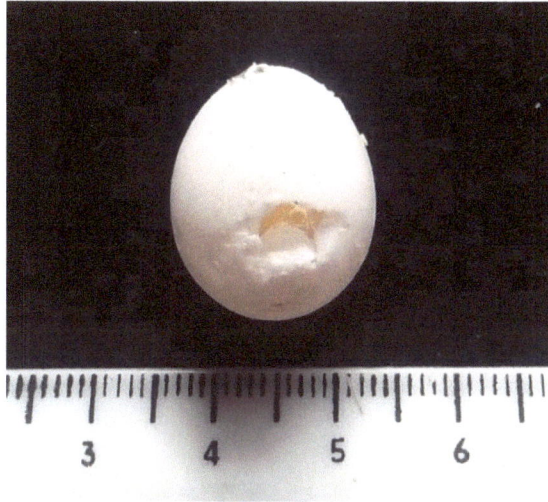

Fig. (2). Pecked Australian parakeet egg.

Fig. (3). Egg of bantam, *Coturnix coturnix*, Agapornis roseicollis and *Casuarius spp.* (from left to right).

INTERNAL EXAM

Before opening, the eggshell should be cleaned with 70% alcohol, and allowed to dry first. The eggshell should be opened by removing a piece along the axis of the shell with an elliptical shape, or by removing a part of the eggshell in the area of the airbag (Kenneth and Rakich, 1994). The puncture is performed with a sharp scissor removing a portion of the eggshell just enough to visualize the contents of the egg. The thickness of the shell and the state of the membranes should be taken into account (by approaching the axis we can obtain a better visualization of the yolk sac and blastocyst), which allows to distinguish the fertilized eggs from the unfertilized one and the position of the embryo (without affecting the yolk sac or the air sac) (Woodford, Keet and Begins, 2000; Schmidt and Reavill, 2003).

The content *in situ* and the stage of development in which it is found are observed. Regarding the content of the egg, alterations of color, size, albumin location, yolk sac, allantoid, presence, and characteristics of the circulatory tree and abnormal odors, should be noted.

In the embryo *in situ*, it is possible to observe the presence of malformations and position changes (mainly the position of the beak and limbs, when the fetus is already in an advanced stage of development), if present. Then remove the embryo and associated tissues. In a small egg, the contents should be poured into a sterile container, such as a petri dish (Fig. **4**). Measure and weigh the embryo, with and without the yolk sac. Measurements of the embryo, in addition to their size, should also be included, when their development permits, the length of the beak, mouth, limbs, tarsus, ulna, tail, *etc.* , depending on the species (bird, reptile, fish or amphibians). An estimate of the age of the embryo is taken, by taking into account the incubation periods of the species. In the case of stillbirth embryos, the dissection should be performed with the aid of a magnifying glass. In the case of small embryos, they can be entirely fixed and later sectioned (Woodford, Keet and Begins, 2000; Schmidt and Reavill, 2003). Developmental abnormalities, such as prognathism or hydrocephalus, should be evaluated. The external examination should include the search for skin lesions and the mouth or nostrils anomalous content. In the incubation muscle, edema, hemorrhage or other injuries could be detected. In the internal examination, the lungs are observed to check the presence of air or inspired fluids. In these cases, histopathology is needed (Fig. **5**).

During the observation of the yolk sac, we can observe, in the case of infections of bacterial origin, lesions such as hemorrhages and coagulation of the material, except in the case of acute death. The yolk sac and its contents should be separated from other tissues and stored separately to prevent contamination of the remaining histopathology samples. Some tissues should be removed for further microbiological, virologic, cytological, toxicological, and other tests. Secondary bacterial infections, with mixed strains, are quite common. If toxicological tests are required, store samples at -20 °C. Exams should include heavy metals, selenium and aromatic hydrocarbons. The remaining tissues should be fixed in 10% buffered formalin for histopathological examination (Schmidt and Reavill, 2003; Butcher and Miles, 2015; Garcês and Pires, 2017).

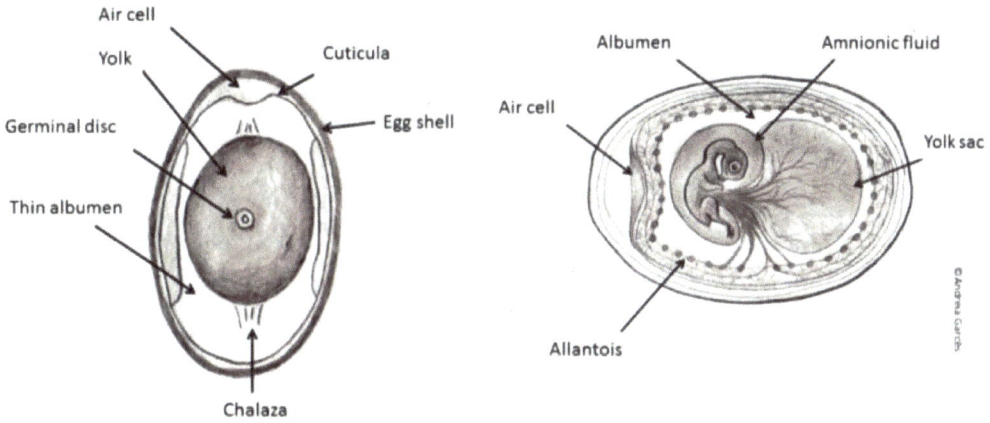

Fig. (4). Scheme of the internal structures of the egg in different stages.

Fig. (5). Opening of the egg and dead embryo.

REFERENCES

Butcher, GD & Miles, RD (2015) *Avian Necropsy Techniques.* Gainesville, Florida.

Chitty, J, Lierz, M (2008). *Manual of raptors, pigeons and waterfowl.* British Small Animal Veterinary Association.

Coles, B (2007) Post-mortem Examination. *Essential of avian medicine and surgery* Blackwell, London 103-15.

Garcês, A & Pires, I (2017) *Manual de Técnicas de Necrópsia em Animais Selvagens.* Arteology, Porto.

Kenneth, L & Rakich, P (1994) Necropsy examination. Avian Medicine: Principles and Application. Wingers Publishing, Florida 355-78.

King, JM (2013) *The necropsy book A Guide for Veterinary Students, Residents, Clinicians, Pathologists, and Biological Researchers.* College of Veterinary Medicine, Cornell University, NY, USA.

King, JM (2014) *The necropsy book A Guide for Veterinary Students, by The Necropsy Book* Charles Louis David DVM Foundation Publisher, Ithaca.

Kinne, J (2015) Post mortem Examination. *Avian Medicine* Mosby, USA 567-81.

Schmidt, RE & Reavill, DR (2003) *A Practitioner's Guide to Avian Necropsy.* Zoological Education Network, Lake Worth, Florida.

Woodford, MH, Keet, DF & Begins, RG (2000) *Post-mortem procedures for wildlife veterinarians and field biologists.* Iucn, Paris, France.

SUBJECT INDEX

A

Abdominal 82, 86, 89, 91, 94, 95
  aortic artery 91
  cavity 82, 86, 89, 94, 95
Abiotic 33
  cadaveric phenomena 33
  phenomena 33
Actin-myosin bond 34
Action 16, 29, 36
  degradative 36
  enzymatic 29
  rapid 16
Activity 29, 34, 111, 131
  enzymatic 29
  metabolic 34
  muscular 34
  reproductive 111
  sexual 131
Adaptations 17, 73, 129
  cardiac 129
  physiologic 17
Adiposity reserves 70
Adjunctive method 11, 13, 14, 16, 17
  of euthanasia 14
  secondary 14
Adrenal glands 53, 55, 62, 63, 90, 111, 120,
    123, 131, 143
Agents 2, 4, 5, 7, 10, 13, 14, 15, 33, 79, 148
  biotic 33
  dissociative 13
  infectious 5, 148
  inhaled 10, 13, 14
  injectable anaesthetic 10
  pathological 2
  zoonotic 79
Air sacs 17, 52, 56, 60, 110, 119, 176
  abdominal 56
  thoracic 52
Alcedo atthis 47, 48, 50, 52
Algor mortis 33, 34
Alligator sinensis 125
Anaemia 51, 78
Anaesthesia 13, 17

Analysis 26
  heavy metal 26
  pesticide 26
Anatomical 91, 102, 163
  differences 91
  diversity 163
  features 102
Anemia 164
Anthropogenic pressure 3
Antigenic stimulation 154
Aquatic 14, 37, 39, 132, 133, 163
  birds 39
  environments 37, 102
  invertebrates 14
  turtles 133
Aromatic hydrocarbons 176
Arteriosclerosis 110, 129
Arthritis 113, 126, 131, 133, 145
Ascaris lumbricoides 7
Ascending distal portion 59
Aspergillus fumigatus 8
Aspiration 12, 24
  section Blood 12
Atherosclerosis 141
Atlantic salmon 165
Autolysis 4, 34, 35, 36, 59, 63, 68, 99, 173
  advanced 63
  post-mortem 34, 35
AVMA guidelines 10, 11

B

Bacillus anthracis 5, 6, 8, 95
Bacteria 5, 6, 8, 24, 35, 36, 60, 124
  pathogenic 124
  psychotropic 124
Bacterial 30, 153
  gastrointestinal load 30
  pericarditis 153
Barbituric acid derivatives 17
Biological communities 31
Biotic cadaveric phenomena 33, 35
Birds 63, 43, 45
  juvenile 63

Ventricle 55, 58, 59, 88, 109, 129, 140, 153
  left 88
Vertebra 12, 45
  cervical 12
Vertebrae 16, 103, 115, 132, 145, 146
  cervical 132
Vertebrate necropsy 158
Vessel telangasia 164
Visceral hypostasis 34

W

Walls 35, 54, 57, 59, 61, 62, 111, 118, 130,
    131, 141, 167
  celomic 118, 130
  chest 61
  dorsal 54, 62
  thin 57, 59
  thoracic 62
  vessel 130
Water 8, 9, 13, 15, 18, 23, 25, 26, 34, 36, 37,
    39, 44, 68, 79, 103, 147, 148, 163, 164
  baths 13
  cold 15, 39, 68
  ice 103, 148
  salinity 37
  shallow 163
  sources 8
  supply 18
  temperature parameters 163
Wildlife 1, 2, 3, 5, 9, 10, 11, 21
  conservation 1
  disease 21
  free-ranging 10
  necropsy 5
  populations 3
  recovery centres 2
  rehabilitation centres 9
  sanctuaries 2
Willis method 25
Wings 32, 51
  hardened 32
  membranous 32
  membranous anterior 32

Y

Yellow fever 6

Z

Zones 37, 73, 131, 149
  lateral 149
  rostral 131

www.ingramcontent.com/pod-product-compliance
Lightning Source LLC
Chambersburg PA
CBHW041702210326
41598CB00007B/504